Biomathematics

Volume 11

B. G. Mirkin S. N. Rodin

Graphs and Genes

With 46 Figures

Translated from the Russian by H. Lynn Beus

Springer-Verlag
Berlin Heidelberg New York Tokyo 1984

Boris G. Mirkin
Central Economic-Math. Institute
of the USSR
Krasikov str. 32
117 418 Moscow, USSR

Sergey N. Rodin
Institute of Cytology and Genetics
630090 Novosibirsk 90, USSR

H. Lynn Beus
TMCB 232, Brigham Young University
Provo, UT 84602, USA

AMS-MOS Classification (1980): 05 C 65, 05 C 75, 92 A 10

ISBN 3-540-12657-0 Springer-Verlag Berlin Heidelberg New York Tokyo
ISBN 0-387-12657-0 Springer-Verlag New York Heidelberg Berlin Tokyo

Library of Congress Cataloging in Publication Data. Mirkin, B. G. (Boris Grigorévich). Graphs and genes. (Biomathematics; v. 11). Translation of: Grafy i geny. Includes indexes. 1. Genetics–Mathematical models. 2. Graph theory.
I. Rodin, Sergeĭ Nikolaevich. II. Title. III. Series.
QH438.4.M3M5713. 1984. 575.1'01'5115. 83-20339.
ISBN 0-387-12657-0 (U.S.)

Printed in Germany

Printing: Beltz Offsetdruck, Hemsbach/Bergstr.
Bookbinding: J. Schäffer OHG, Grünstadt
2141/3140-543210

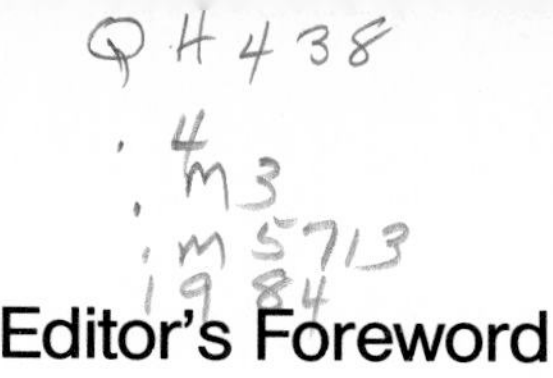

Editor's Foreword

This book is written by a mathematician and a theoretical biologist who have arrived at a good mutual understanding and a well worked-out common notation. The reader need hardly be convinced of the necessity of such a mutual understanding, not only for the two investigators, but also for the sciences they represent. Like Molière's hero, geneticists are gradually beginning to understand that, unknowingly, they have been speaking in the language of cybernetics. Mathematicians are unexpectedly discovering that many past and present problems and methods of genetics can be naturally formulated in the language of graph theory. In this way a powerful abstract mathematical theory suddenly finds a productive application. Moreover, in its turn, such an application begins to "feed" the mathematical theory by presenting it with a number of new problems. The reader may judge for himself the fruitfulness of such mutual interaction.

At the same time several important circumstances need to be mentioned. The formalization and rigorous formulation given here embraces not only the older problems, known by geneticists for many decades (the construction of genetic maps, the analysis of complementation, etc.), but also comparatively new problems: the construction of partial complementation maps, phylogenetic trees of proteins, etc. Furthermore, the formalization process, of necessity, requires the working out of a rigorous system of genetic understanding, and also the formulation and solution of a number of problems which have not attracted attention in the theory, up to this time: the problem of the uniqueness of genetic mapping, the search for a best approximation to a map, problems of invariants, etc.

It is interesting that genetics, as a field of application for graph theory, turns out to be such a rich one. Genetic methods produce information about genes, proteins and genetic systems in the form of a great variety of partial manifestations obtained through crosses. The theoretic task of reconstructing, from a great mass of data, the condensed image of a system is represented in graph theory either as the problem of representing matrices by graphs or as a problem of approximation.

It is important to emphasize that the methods of graph theory are not only useful in solving many important genetics problems, but in some cases, lead the investigator into the sphere of more fundamental problems, such as determining fine internal gene structure, studying the "semantics" of the genetic language, the genetic organization of ontogenesis, the "nonadaptive" evolution of macromolecules, etc. In this sense the attempt of the authors to unite graph theory with the theories concerning genes and genetic control systems may already be considered successful. We wish them further success in their endeavor.

Prof. V.A. Ratner

Preface to the English Edition

Mathematical genetics has traditionally been concerned with the growth of genetic systems as populations, utilizing as a basis the theory of determinate and stochastic differential equations. However, there exist a number of important areas of genetics which do not lend themselves to this approach. The genetic problems to which this book is devoted -- the analysis of basic principles of organization, functioning and evolution of genetic systems on the molecular level -- are quite adequately handled by another branch of mathematics: graph theory. This is only natural, when one considers the discreteness and quite rigid structural nature of many functions of molecular-genetic control systems.

Since the moment in 1976 when work was completed on the Russian version of the book there have been a number of changes in theories about genes, some of them being very significant. These became possible due to the rapid growth of the technology of direct biochemical "reading" of deoxyribonucleic genetic texts. And since this book is devoted to basic problems in deciphering genetic texts, one might think that its material has become significantly outdated.

But this is not so. The fact is that biochemists are deciphering only the "passive" information contained in DNA. The activation, the allocation of this information according to meaning, can only be achieved (at least, presently) by the application of independent genetic data about the interactions of different variants of some genome fragments or other. But for such an "indirect" approach the language and methods of graph theory turn out to be extremely useful. For this reason the contents of the book are practically invariant under the accumulation of new information about the structure of molecular-genetic systems.

Thus in the preparation of this edition we limited ourselves to the minimal additions and modifications necessary to sharpen positions and results in light of the latest conceptions. The single exception is the material of chapter three, which was reworked and enlarged on the basis of the appearance of recent, more subtle approaches to the analysis of organism evolution at the molecular level (section 3.1.5 was added, and section 3.2.2 was rewritten).

Novosibirsk | B. G. Mirkin
14 Jan. 1982 | S. N. Rodin

Preface to the Russian Edition

No more surprising to us is the monstrous whale,
Than the detailed parts of a tiny worm or snail.
Great in heaven is the creator of all these things.
His greatness in earth the tiniest of all these sings.

-- M. V. Lomonosov
Epistle on the use of glass

As is well known, the theory of graphs provides a universal language for representing the structural organization of systems of arbitrary nature. Its use in describing the organization of genetic systems is of special interest, since their functioning is fundamentally determined by their structure.

In the last ten to twenty years the growth of molecular biology has elicited the fact that all basic characteristics of living systems are determined, finally, by linearly-written, discrete, hereditary information contained in the special class of macromolecules called DNA, and by a complex system for reading out this information and recoding it in the language of protein macromolecules, the activity of these being determined by their spatial structure [31]. It is quite natural to use the language of graph theory in studying both the fundamental linear genetic texts and the three-dimensional protein structures which implement DNA instructions. It is especially important in cases when direct biophysical investigations become difficult or impossible, since genetic methods based on the crossing of mutant forms have high resolving ability (at least for microorganisms). Genetic methods provide only indirect information, with graph theory serving as an instrument to draw from such information a "portrait" of the underlying genetic systems.

Recently, both in the literature of mathematics and genetics, articles along this line of thought have appeared more and more frequently. This monograph attempts (through graph theory) to systematize results obtained from the study of the structure, functions and evolution of genetic systems.

Genetic problems have led to new ideas and new problems in graph theory. In turn, mathematical methods have led to more precise descriptions of these same genetic objects and, as a result, to a more advanced understanding of how molecular-genetic control systems function. We emphasize that although independent results, both about graphs and about genes are presented in the book, the essence of our work lies in their union -- in demonstrating the fruitful applications of these mathematical methods in the investigation of specific genetic systems.

There are two stages in the application of mathematical methods to real data. The first consists in forming a mathematical model of the phenomenon and investigating its formal character. As a rule, experimental data are not represented accurately enough by the formal model, since it is impossible to cover all particular real situations. From this comes the necessity of a second stage: the construction of a mathematical theory of the transition from real data to constructed models. These two stages -- the modeling of phenomena and the processing of data -- appear in one form or another in any applied mathematical investigation.

For our purposes in this book, the first stage consists in modeling the structure of genetic objects of interest and in describing the properties of the data which may be represented by such a model. The essence of this stage lies in *representing* empirical data by ideally constructed models. The second stage consists in *approximating* experimental data, which are complex in structure, by models of more simple structure. Consequently, this book is concerned with two types of mathematical prpblems: the representation of structures, and the approximation of structures.

From this point of view we will now describe the mathematical nature of the book.

In the first chapter we investigate a model of the arrangement of mutational defects in a chromosome, i.e. a linear map. This model is simple enough to allow mathematical investigation, yet complex enough to not be trivial. The basic problem here is one of representation, i.e. the description of those data structures for which the linear map model is valid, namely, graphs and hypergraphs of intervals.

A considerable number of interesting mathematical results have accumulated in the theory of graphs and hypergraphs of intervals, and their presentation requires considerable space. At the same time, a general theory of the approximation of data by such devices does not yet exist. Consequently, in concrete examples the necessary approximation was done "by eye", since in these cases the structure of the data was either sufficiently close to the theoretical, as in Sections 1.5.1 and 1.5.3, or was so distant that it led to rejection of the model (Section 1.5.2).

The mathematical object of Chapter 2, the concept of structure, is extremely simple, and its mathematical theory is trivial. Consequently it was possible to focus on questions of approximation. The basic mathematical content of the chapter (Section 2.2) consists of the description of models and of methods of approximating data by structures of a special form. These models constitute part of the general theory of qualitative data analysis [22]. The complexity of the approximation tasks examined is considerable: These are the so-called NP-complete [74a] problems of discrete optimization, and precisely because of this their solution by mathematical means is hardly possible. However, as our attempt shows, the algorithms developed for the localized approximation of specific data lead to adequate results (and not only for genetics).

In the third chapter the tasks of approximation and representation are relatively balanced. The basic model is the notion of a weighted tree, admitting a not-too-trivial description. A theory of approximation of data by means of trees has not yet been formulated. We examine mainly those "working" algorithms which are useful in obtaining the genetic results of concern. In our view the concept of a weighted tree, because of its simplicity, should attract the attention of mathematicians toward working out a corresponding theory of approximation.

It must be said that the mathematical concepts of graphs and hypergraphs of intervals, of structures, and of trees, examined in this monograph are extremely universal and applicable. Thus graphs of intervals find application in problems of archeology and circuit design; methods of approximating graphs by structures, in the study of structural organiation and in classification problems; trees, in the organization of information systems, etc. This means the mathematical content of the book is of use not only for genetics applications.

We will now describe the genetic content of the book. Experimental genetic data are derived directly from interactions of nucleotide texts in crosses (recombination analysis), or more indirectly from information about the interaction of protein products of genetic systems (complementation analysis). Most genetic analyses are based on information of the first kind. Using such data, it is not difficult to construct sufficiently complete and reliable portraits of genes as to be useful in various further applications. "Functional" data about proteins are seldom used. This is explained primarily by the indirect connection between the interactions of proteins and the interactions of their parent nucleotide texts.

The methods described in this book are effective precisely for such indirect information, where the usual methods of genetic mapping "do not work." In this book, then, our primary attention is focused on the analysis of information about the interactions of protein products. It is the authors' hope that the methods presented here will indicate new possibilities in complementation analysis.

In Chapter one, following presentation of the basic ideas about genetic control systems, we show how genetic data can be used in studying the organization of chromosomes. In the second chapter some highly controversial questions are discussed, dealing with the description of the functional organization of protein macromolecules. New concepts developed by V. A. Ratner and his collaborators are described, and corroborative analysis of experimental data is carried out. In the third chapter data about the structures of the proteins of a hemoglobin family are used in analyzing its evolution. Qualitative analysis of the evolutionary tree produced by this process leads to a number of important conclusions about the course of evolution.

The authors have tried to write so that the book can be read relatively independently of other sources. The necessary ideas of genetics are introduced in Section 1.1, and those of graph theory are given in an appendix. For the benefit of the reader an index of genetic terms is provided, and also an index of mathematical terms. The usual method of indices is used for internal references, including the number of the chapter and section which contain a formula or subsection.

A significant portion of the material of the book reflects the authors' own investigations, or more accurately, the investigations of the informal collective to which they belong. In this regard the authors express appreciation to all the members of the collective. We especially wish to mention the role of V. A. Ratner, not only as editor of the book, but also as leader of and active participant in all of the new genetic investigations described in it. His ideas on molecular-genetic control systems serve as pithy fundamentals to the book. Many new mathematical results were obtained in conjunction with V. L. Cooperstokh and V. A. Trophimov, who also carried out a significant portion of the machine computations. In particular, the contents of the second chapter are based to a significant degree on their work. The helpful criticisms of Yu.M. Svirežev and I. B. Muchnik significantly aided in improving the structure of the book.

Table of Contents

Chapter 1. Graphs in the Analysis of Gene Structure

§1. Gene systems and their maps

1. Levels of the genetic language. The genetic apparatus of living things stores hereditary information which ensures the reproduction of the organism. Another of its purposes, no less important, is to provide for normal functioning of each cell of the organism in the process of its individual growth (*ontogenesis*). This means that the hereditary memory of a cell (*genome*) governs all aspects of its vital activities, and the hereditary memory of the organism, in the final analysis, governs its functioning and reproduction as a whole, through the mutual interactions of cells.

The genetic program is concentrated in the chromosomes of the cell nucleus and is written down in the thread-like polynucleotide molecules of DNA -- *deoxyribonucleic acid*. The individual nucleotides, or more precisely, the nitrogenous bases contained in them, may be considered as characters of the "text" of a program. These *nitrogenous bases* (characters) are of four types: *adenine*(A), *thymine*(T), *guanine*(G) and *cytosine*(C).

Thus the primary governing information is represented as a linear record in this four-letter alphabet.

This information is used by the cell to synthesize needed proteins, and also in carrying out some other functions. It is the *proteins* which perform the basic biochemical operations in a cell, that maintain its vital activities. As a rule each protein performs one (or in some cases, a few) specific biochemical operation: catalysis of a biochemical reaction, transport of material, molecular control, etc. Therefore, in the cells of living organisms many thousands of molecules of various proteins are continuously being synthesized. A single protein molecule appears as a complex twisted *polypeptide chain* -- a sequence of hundreds of *amino acids* of twenty different kinds. In other words, a polypeptide chain is a text in a twenty-character alphabet of amino acids.

The translation of the genetic content from the language of nucleotides to the language of amino acids proceeds as follows. Each amino acid is coded by a definite ordered triplet (*codon*) of nucleotides, so that an arbitrary message in DNA is transferred by this *triplet code* into a sequence of amino acids (a protein).

Some redundancy is evident in the fact that 64 codons are used for 20 amino acids (Table 1). However, there are not 44 seemingly "meaningless" triplets, in terms of amino acids, as would be expected in a one-to-one coding, but only three. Each of a number of the amino acids is coded for by several triplets. The three "meaningless" codons play the role of punctuation marks in the genetic text. In Table 1 U is used in place of T ([31], p. 4).

The question arises, what is it that requires this representation of instructions (for each function) in two notations: the one in the language of nucleotides (for the synthesis of protein) and the other in the language of amino acids (for the direct fulfillment of this function). It seems evident that all of the necessary instructions could be written in only one language. In such a system there would be no problem of translation from one language to the other (about which, incidentally, something is said in this book).

Before answering this question we must consider several chemical properties of nucleotides and amino acids. In most cases a *molecule* of DNA consists of two coupled strands of nucleotides (the double helix of Watson and Crick). The nucleotides of one strand correspond chemically in a one-to-one way with those of the other, according to the rule: A always corresponds with T, and G with C (and the reverse). On the one hand, this makes a molecule of DNA stable under various influences, and on the other hand, it assures exact copying in each *replication*. In the replication process the helix of DNA unwinds, and on each of the strands in turn a new strand is synthesized, according to the pairing rules given above. As a result two daughter molecules are obtained, each having the same text.

At the same time, the stated characteristics of nucleotides (primarily paired interactions) are not sufficient to produce the variety of functions necessary in maintaining the vital activities of a cell.

We now consider proteins, whose spatial structure is entirely different. The initial sequence of amino acids forms the so-called *primary structure*. Due to the physico-chemical interactions of its component amino acids the primary structure sometimes acquires an extremely convoluted spatial form. The intermediate stage of this packing is called the *secondary structure*, and the final result, the *tertiary structure* of a protein. Packing does not stop with this third stage. Frequently, separate polypeptide chains unite in an integral complex called the *quaternary structure* of a protein (a *multimer*). On the tertiary and quaternary levels special sections of spatially-close amino acids known

The Genetic Code [31]

First nucleotide codon	Second nucleotide codon				Third nucleotide codon
	U	C	A	G	
U	phe	ser	tyr	cys	U
	phe	set	tyr	cys	C
	leu	set	Term.*	Term.	A
	leu	set	Term.	trp	G
C	leu	pro	his	arg	U
	leu	pro	his	arg	C
	leu	pro	gln	arg	A
	leu	pro	gln	arg	G
A	ile	thr	asn	set	U
	ile	thr	asn	ser	C
	ile	thr	lys	arg	A
	met+	thr	lys	arg	G
G	val	ala	asp	gly	U
	val	ala	asp	gly	C
	val	ala	glu	gly	A
	val+	ala	glu	gly	G

Note: Amino acids are given in their usual three-letter abbreviations (phe - Phenylalanine, leu - Leucine etc.).

+ Codon initiators (at the beginning of translation)
* Codon terminators (at the end of translation)

Table 1

as *functional centers* directly carry out the biochemical operations characteristic of a given protein. It is important to note that functional centers are conglomerations of various amino acids, so that their action is not reducible to the characteristics of the individual components.

However, amino acids do not display any kind of pronounced one-to-one pairings. Occasionally, it is true that fragments of polynucleotide chains form structures (so-called β-structures) which outwardly remind one of pair-wise-connected polynucleotides. However, this is exhibited in extremely specific localizations of amino acids, and is not characteristic of them as such [31,52]. This leads to the reverse consequence, in comparison with nucleotides. Protein molecules cannot reproduce themselves by template synthesis†; on the other hand, due to the associated packings of specific amino acid sequences, they are able to realize, in practice, any molecular function required by the cell.

Thus the existence of two languages means that there is a separation of responsibility: DNA is the repository of hereditary information (the hereditary memory of the cell), while proteins are the functional structures. This separation leads to high noise stability in genetic information. Having separated the hereditary memory of a cell (DNA) from participation in the indirect aspect of cell functioning, nature provided for its protection from various external influences. It "hid" *chromosomes* (structures which contain DNA) in the nucleus of the cell, while the synthesis of proteins and their succeeding reactions were removed primarily to the cytoplasm.

Such a spatial separation necessarily implies a "go-between", a messenger which carries information from DNA to the places of protein synthesis. This role is played by *messenger ribonucleic acid* (mRNA). As with DNA, it consists of four nucleotides: *adenine*(A), *quanine*(G), *cytosine*(C) and *uracil*(U). Although uracil(U) is used in place of thymine(T), specific pairings of associated nucleotides are maintained also in this case, namely A with U and G with C. In accord with this rule mRNA is synthesized on one of the strands of DNA. This process, analogous to the reading-out of information, is called *transcription*. Then the mRNA moves into the cytoplasm and joins itself to a *ribosome*, a special organelle of the cell on which protein synthesis takes place. Beforehand, amino acids are bound to molecules of RNA of a second type

---

†More precisely, this is not only a matter of the absence of specific paired recognition of amino acids. M. Aigen showed convincingly the basic possibility of non-templace reproduction in so-called autocatalytic cycles. However these systems are not able to evolve, i.e. to reproduce changes (mutations) arising from mistakes in copying. For more details see [52].

called *transfer RNA* (tRNA), and in this form they approach the ribosomes "loaded" with mRNA. In accordance with the genetic code each amino acid is bound to a specific tRNA. A molecule of tRNA contains a particular triplet, known as an *anticodon*, which recognizes its codon in mRNA. Hence, an amino acid "riding" on tRNA is included in the synthesized polypeptide chain at the required position so that the collinearity of the chain and mRNA is ensured. This process of transferring information from mRNA to the language of polypeptides is called *translation*.

A chromosome contains information for the synthesis of not just one, but many different proteins (as many as several thousand). Even for a relatively simple organism such as the bacteriophage, a chromosome controls the synthesis of more than 50 different proteins. Portions of the text corresponding to separate proteins are called *cistrons*. Cistrons are separated from one another by special "punctuating" nucleotides: *Translation initiators* are AUG and (sometimes) GUG, which correspond respectively to the amino acids methionine and valine. *Translation terminators* are the "nonsense" codons UAA, UAG and UGA. Thus the cistron is the unit of translation. The basic action of a cistron is expressed in the functioning of a specific protein, and the phenotypic traits of an organism depend, in the final analysis, on just such protein functions. It is known that a living organism is an integral system, with each cistron influencing components of the system in one way or another. However, one can always identify one or two traits whose formation (by means of corresponding proteins) is critically dependent on a particular cistron. In this sense we say the cistron *controls* the given trait.

The term *cistron*, which arose in molecular genetics, is essentially the successor of the classical term *gene*. The historical conception of a gene arose in the specific context when individual discrete genes were associated with autonomous traits.

It should be noted that the cooperative actions of several proteins may be necessary to ensure certain functions and to form the corresponding trait. In many cases a "complex" function in a microorganism corresponds to a group of cistrons working in coordination. Typical of systems of this kind are the units of transcription, called *scriptons*. At the level of DNA, a scripton contains sequences which mark the beginning (*promotor*) and ending (*terminator*) of transcription, while the separation of this single text into cistrons corresponding to individual polypeptide chains is performed on mRNA in the translation process.

What is the physical meaning of the union of an entire group of cistrons under transcription? It turns out that the protein products of the cistrons contained in a scripton do not function independently, rather they control interconnected stages of the transformation (usually synthesis or decomposition) of materials within the cell, and frequently enter into common multimers as well.

Consider, for example, the breaking-down within the cell of some energetic raw materials such as carbohydrates or amino acids for the extraction and storing of energy. Here the associated cistrons must be switched off when the raw material is running low, and switched on when it is sufficient. The switching of a cistron is performed by a special control segment called an *operator*, situated at the beginning of the scripton. Scriptons with operators are called *operons*, and work on the principle of a control system with feedback. The operator of an operon is sensitive to a special protein, which is synthesized under control of still another special gene, a *regulator*. In the absence of needed raw material this protein (called a *repressor*) represses the action of the operon by combining with the operator and preventing transcription of the entire remaining portion (the "barrier" is closed). If there is sufficient raw material present it combines with the protein repressor and changes its spatial configuration so that it no longer recognizes the operator (the "barrier" is open). Then the operon synthesizes proteins (enzymes) which break up the raw material until its concentration falls below the level necessary for it to bind the operon repressor, and consequently the operon is again turned off.

Such control systems have been found, up to this time, only in microorganisms. Systems have now been described consisting of many operons and scriptons, and recently, in the case of the λ bacteriophage, the entire control system for individual growth has been deciphered.

In higher organisms it has been impossible to demonstrate the presence of operon-like systems (see however Section 5.3). The construction of the cells of such organisms permits a wide spectrum of possibilities for the organization of regulation, unthinkable in the lower organisms. These possibilities are connected with the spatial and temporal disconnection of the processes of transcription (the synthesis of messenger RNA in the nucleus) and translation (the synthesis of polypeptides in the cytoplasm) and to all appearances are realized through the genetic redundancy of the corresponding macromolecules of DNA and RNA. The initial transcript synthesized in the nucleus is, as a rule, much longer than the mRNA which is translated into a protein: biologists say a maturing or processing RNA takes place.

One kind of redundancy was recently deciphered. By means of direct biochemical investigation of DNA sequences in higher organisms it was revealed that a significant portion of the initial transcript of a cistron is not translated into polypeptide text. Such a cistron is divided into regions of two types: *exon* and *intron*. The exons are translated, while the introns are not. The noncoding intron fragments sometimes reach significant total length. Thus, the noncoding part of the β-cistron of mammalian hemoglobin makes up on the order ot two-thirds of its length (for more details see page 154).

The discovery of introns in 1977 gave rise to a number of serious questions concerning their roles in internal cell processes. The picture is still not clear. It can be supposed only that the intron organization of genes provides the possibility for additional means of regulating gene activity and heightens their adaptive potential. For the genes which maintain active molecular immunity in higher organisms, for example, mammals, this thesis has been demonstrated and can be given in detail.

The characteristics of the immune system's reaction to foreign substances -- *antigens*, found in the organism, are determined at the molecular level by the formation of corresponding complex protein macromolecules -- *antibodies*, or *immunoglobulins*. The diversity of the antibodies must be very great, in view of the practically unending number of different antigens. The general outline of the mechanism for generating such diversity is as follows. Initially in the genome there is a significant quantity of cistrons, coding for so-called *domains* -- spatial and functional autonomous blocks for the formation of immunoglobulin molecules. These cistrons are localized in the form of clusters, gene batteries, but separated from one another by intercistronic nucleotide sequences called *spacers*. One theory says that in response to an antigen a selection is made in the genome among all the possible combinations of these cistrons, combining them into a single DNA sequence. This selection comes about through a special recombination enzyme system, which recognized spacers and combines separate cistrons. A variety of immunoglobulins are produced by translation of the DNA sequences thus formed. The cells synthesizing an immunoglobulin necessary to inactivate a given antigen gain an advantage in propagation in order to maintain the immunity. Direct biochemical analysis (sequencing) of the genes which code for immunoglobulins, has shown that exons correspond to domains in these proteins. This permits the supposition that the introns are former spacers [66a]. On the basis of this example one can clearly see how, in principle, the intron structure can be utilized for the goals of evolution and regulation.

In wide circulation also is the idea that since without the excision of introns and some other fragments of mRNA synthesis of a single polypeptide sequence is not possible, the presence of introns may be used by the cell for the control of gene activity at the level of translation.

Thus within the genetic language there are several identifiable layers.

The "upper" layer is the language of DNA, the hereditary memory of the cell. The meanings of texts coded in the four-letter alphabet of DNA are completely unambiguous. Each word (codon) is rewritten as exactly the same word (codon) of mRNA, except that the letter T (thymine) is replaced by the letter U (uracil).

Texts in the alphabet of mRNA are also unambiguous in meaning: Each codon contains an instruction for the inclusion of a specific amino acid (in agreement with the genetic code) in the associated polypeptide chain, the "phrase" (cistron) being the instructions for synthesis of the entire chain.

Thus the third layer in the genetic language is the primary structure of proteins, i.e. linear texts in a twenty-character alphabet of amino acids. Contained in these texts are the instructions for the necessary spatial packing of polypeptides in active protein molecules. Unlike the *genetic code* (the transition from mRNA to the primary structure of proteins), these instructions (for the transition from primary to tertiary structure, i.e. for the self-organization of protein molecules) have not been completely deciphered, although recently more papers have been appearing which report progress in the solution of this complex problem.

The next layer of the genetic language is the three-dimensional structure of protein molecules. At this level, segments of proximal amino acids form *functional centers* containing instructions about "what to make" (catalytic activity) or "when and how to act" (regulatory function) or "where to go" (intracellular organization), etc., in which functions are performed as prescribed by genetic texts of the "higher" layers. Little is known about the alphabet, much less the grammar, of the language of functional centers (see page 93).

The language of the next layer is that of the functions carried out by proteins. It has had little investigation, although recently it became clear that the "dictionary" of this language is not so large as one might expect. For example, all *proteolytic enzymes*, which decompose "edible" protein molecules, act according to the same molecular scheme, differing only with respect to which adjacent amino acids are recognized as a point of cleavage.

In this way each layer of the genetic language is distinquished from the next by the alphabet and syntax of its texts, and by its *seman-*

tics, which are realized in the synthesis of texts of the next layer, and finally, in the vital activities of the cell determined by the functioning of protein molecules.

In moving through the layers of the genetic language (from DNA to the functional centers of proteins), the semantics of the genetic information becomes more and more "active." At the same time, the active form of the language of the protein layer, though structurally stable, is quite flexible. It is true that the initial polynucleotide text (in DNA) uniquely determines the molecular function of the corresponding protein. This does not mean, however, that the function of a protein requires a unique primary structure. There are key positions within the primary structure of a protein which determine its spatial packing, and consequently, its function. Thus a particular molecular function may correspond to a whole class of polypeptides, which differ only with respect to the amino acids occurring in the positions of less functional importance, the non-key positions. In other words, this spatial method of realizing the semantics of the "protein" layers allows the existence of synonyms. Actually, synonymous polypeptides do provide for some nuances of meaning within their instructions. Without the existence of such synonyms it is difficult to imagine a process of change and evolution in genetic (or any developed) systems [31,32,52]. The synonymity of the genetic language (as also the usual "human" languages) ensures the capacity for flexible response to changing environmental conditions without the creation of fundamentally new texts, by arranging (within the hereditary memory (DNA)) for only local (non-key-position) changes in the primary structure of proteins. Roughly speaking, associated with the "continuity" of change in the external surroundings there must be a "continuity" of change in the meaning of the significant constructs of the language. Thus, synonymity is characterized not only by a multiplicity of meanings but also by the blurred nature of the set of semantic values.

The transitions from layer to layer of the genetic language are used in regulating and controlling the realization of genetic information. The process of transcription (the transition from DNA to mRNA) is, evidently, central to the regulation of gene activity in such simply constructed organisms as bacteria and viruses. In higher forms the control possibilities are much broader. As an example of this, we note the system of control for the translation process (the transition from mRNA to protein) through the excision of introns. The greater the descent from the higher layers, the more restricted becomes the sphere of action of control (if elements of the control mechanism itself are not affected).

Languages of human intercourse differ significantly from the genetic language in that they are broader and richer, being much more than languages of instructions (commands) [31]. In this sense the genetic language resembles more closely the languages of socio-economic control systems, which are inherently flexible and hierarchic in nature. The comparison of these languages would probably give a deeper understanding of the general principles of organization of control in complex systems, but we know of no interesting progress in this area. It is accepted that there are essential differences between the genetic and the socio-economic control systems in the "development" of their administrative instructions. No matter how many layers are found in the genetic language, it must always be remembered that all of the necessary information is contained (in a "passive" form) in the initial polynucleotide text. Successive stages simply recode it into active "working" structures. In socio-economic control systems instructions arriving at their level of application may be noticeably changed due to conscious "creative" effects at each preceding level. However, even in genetic systems the process of recoding is not always so faultlessly exact as might be expected. In fact, every genetic control is, in one way or another, connected with a change in informational macromolecules and with corresponding modifications in their structure.

The properties of the genetic language make it similar to the languages of programming systems, which also consist of instructions, and are many-layered. Texts in a high-level programming language (the upper layer) are translated into texts of internal machine langauges (the langauge of micro commands and the language of physical processes necessary for computation), which ensure the completion of the initial prescriptions.

Considering the *collinearity* of the polynucleotide and polypeptide forms of recording genetic information (excluding introns), we will differentiate (conditionally, of course) within the genetic language its *structural* and *semantic* levels. The first level is that of the linear (collinear between them) polynucleotide (DNA, RNA) and polypeptide texts, while the second level consists of the functional centers of protein macromolecules, i.e. the level of "active" genetic information. We will keep in mind, though, that such an important stage in the realization of the semantics as the separation of one instruction from another, is effected in the polynucleotide layer (of DNA,RNA) by means of special punctuation marks [31]. It should be particularly emphasized that the differentiation within the genetic language of the levels of reproduction (DNA) and the functioning (proteins) greatly simplifies the real picture of the interactions between nucleic acids

acids and proteins. The fact is that in living cells the basic processes of recoding -- replication, transcription and translation -- can take place only with the participation of special enzymes. The hereditary memory of a cell (DNA) must, of necessity, contain instructions for the synthesis of the entire set of macromolecules (enzymes, tRNA and others) which "serve" these processes. In other words, the mechanism of self-reproduction in real cells will not work in the absence of the products it itself produces [31,52]. For this reason M. Aigen correctly remarked that even in genetic systems of relatively simple construction (viruses, bacteria) the general scheme of inter-relations among nucleic acids and proteins appears as a complex hierarchy of "closed loops" [52].

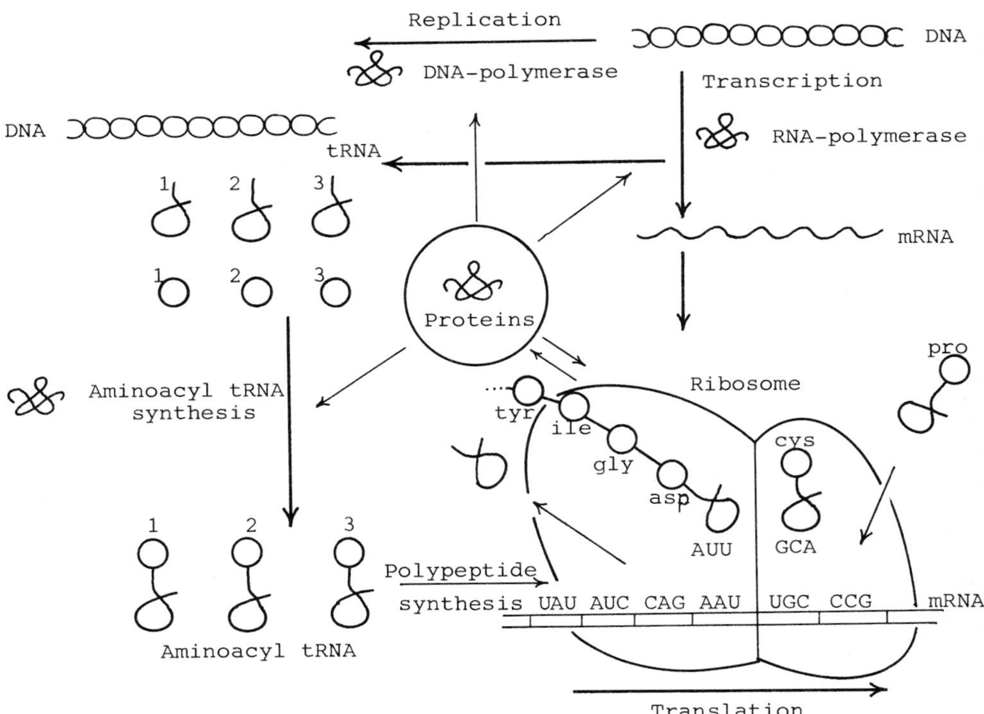

Figure 1. Inter-relations of nucleic acids and proteins in the process of their reproduction [52].

Fig. 1 shows the fundamental functional connections of a genetic system [52]:

1. DNA, on which is carried out the template synthesis of complementary chains by the enzyme DNA-polymerase.

2. mRNA being synthesized on DNA by the enzyme RNA-polymerase.

3. tRNA and aminoacyl-tRNA synthetases. These enzymes connect molecules of tRNA with certain amino acids, thus ensuring a correspondence between "words" of the polynucleotide and polypeptide texts.

4. A ribosome -- an enzymatic complex, including molecules of ribosomal RNA. A ribosome consists of two unequal subunits: In the smaller is located the acceptor segment (the center of connection with tRNA), and in the larger there is a catalytic center for forming peptide connections with neighboring amino acids. At each moment during translation there are two molecules of aminoacyl-tRNA associated with the ribosome, the anticodons of which are complementary to the corresponding codons of mRNA. In the scheme of Fig. 1 is shown one of the stages of movement of a ribosome along mRNA: The amino acid cysteine is ready to join with an already-translated polypeptide fragment (...—tyrosine—isoleucine—glycine— asparagine—).

5. The proteins appear in the center of the scheme: on the one hand, as end products of processes of recoding (thick arrows), and on the other hand, as necessary participants in these same processes (thin arrows).

In the sequel we will use the term *gene* in the broad, although inexact, classical sense, to designate the unit which controls a separate trait, keeping in mind that such a gene may consist of a single cistron or of multiple cistrons. This is important in view of the fact that in genetic experiments the investigator works, as a rule, not with molecular objects, but with the mutant variants of the trait of interest.

2. Mutations, recombination, complementation. Without the process of genetic reproduction described above, the basis of which is the template copying of molecules of DNA, neither growth nor reproduction of multicelled organisms would be thinkable. The process of growth is, essentially, a series of successive cell divisions, the first of which is the division of the embryonic cell, or *zygote*. The basic hereditary material of the zygote is concentrated in the nucleus in the form of a fixed number of discrete structures -- the *chromosomes*. Prior to cell division, each chromosome is doubled (by doubling DNA), and during division of the nucleus identical copies of the chromosomes are separated equally to each daughter cell. This cell division process, called *mitosis*, ensures the transfer of a full complement of chromosomes to

each daughter cell.

The basis of sexual reproduction is an essentially different process of cell division called *meiosis*. In the formation of a male or female sex cell (*gamete*) the parent cell divides twice, so that in the daughter cells there is only half the usual number of chromosomes. However, this entails no loss of genetic information. The fact is that in the cells of higher plants and animals each type of chromosome appears twice, as *homologues*. Such chromosomes are said to be *homologous*, and such cells are called *diploids*. In the products of meiosis (the *gametes*: ovum and spermatozoid) there is only a single copy of each chromosome. Such cells are called *haploids*. In the process of fertilization the male and female haploid gametes combine to form a diploid (*zygote*), re-establishing a normal diploid set of chromosomes, in which each gene again appears twice.

The genetic material of bacteria and a number of other single-celled organisms is represented by a single molecule of DNA, i.e. such organisms are typically haploids. The cell division process differs here somewhat from that described above though it retains the same purpose: to transfer a full complement of the DNA to each daughter organism.

In DNA replication occasionally there are "misfires", mistakes in reproduction, called *mutations*. The majority of mutations are the result of a simple replacement of individual nucleotides (characters). If, as a result, there is a replacement of amino acids in a protein we speak of a *missense mutation*; if the new triplet turns out to be meaningless then such a replacement is called a *nonsense mutation*. More complex mistakes are possible: the loss or insertion of separate nucleotides (such mutations lead to a displacement of the "read out frame" and the distortion of the entire subsequent text), loss of an entire DNA fragment (*deletion*), *duplication* of DNA segments, etc. The fact that all of these mistakes are reproduced in the replicating DNA means they are passed on to descendants, i.e. inherited.

Since the translation of separate genes (cistrons) is independent, mutations internal to a gene usually distort the function of only the protein it controls, and that locally.

Among all the possible nucleotide replacements a significant portion are *synonomous*. The genetic code is such that replacement of a single nucleotide in a triplet frequently produces a codon for exactly the same amino acid. There are also other, more subtle regulating properties of the genetic code and the translation apparatus which increase the noise stability of the genetic texts: Codons with similar sequences code for amino acids with similar characteristics; amino acids are incorporated more readily into protein structures, the more triplets there are that correspond to them, etc. (see Table 1 and [2,31,52]). However, the noise stability thus attained has its limits. It can be shown [2,31] that if the chromosomes of haploid descendants have a total length greater than $10^7 - 10^8$ nucleotides, then they will inevitably carry at least one *lethal* (death-causing) *mutation* (called a *lethal*).

This establishes a limit to the complexity of the organization of living systems at the haploid level (actually, existing haploid organisms have fewer than 10^7 nucleotides [2,31]). We cannot rule out the possibility that this may be the reason for the transition from a haploid to a diploid structure in the hereditary mechanism, for in the diploid system all the chromosomes are doubled, so that each gene has two representatives. Of course, in the cells of multicellular organisms there are several such pairs of homologous chromosomes (for example, the fly *Drosophila melanogaster* has 4 pairs, humans have 23, maize 10). If only one of the homologous chromosomes is mutated then, in the presence of a normal partner, its defect is not usually manifest. We will return to a more precise discussion of this question later.

Generally speaking, doubling of information may also be realized by another means: the writing down, in order, of the same text twice in a single "carrier." Some genes are repeated exactly this way. However, the doubling of hereditary information on different but homologous chromosomes not only increases its noise stability, but also leads to a fundamental increase in evolutionary possibilities. These possibilities are realized in higher organisms specifically through the process of sexual propagation described previously, which is connected with just such an information-doubling system. Indeed, in organisms propagating by sexual means, each chromosome of a homologous pair has a paternal or maternal origin, and during meiosis members of each pair are distributed to gametes randomly. As a result a random combination of chromosom pairs is produced, which in its turn leads to the appearance of mixed descendants. Thereby, the genetic diversity of that particular species of organism is significantly increased, and the possibility for the selection of new forms becomes greater.

However, an increase in genetic diversity is attained not only by the exchange of whole chromosomes, in which the genes contained in one chromosome are inherited as a group (*cohesively*). There is also a process of intrachromosomal *recombination* of genes called *crossing-over*. At one stage of meiosis homologous chromosomes draw close together collinearly (*synapse*) and exchange their homologous fragments. By this process exchange of paternal and maternal genes can occur. If, for example, homologous chromosomes differ by two mutations, only one of these being contained in each of the exchanged fragments, then as a result of crossing-over the chromosomes will be recombinant.

The vital feature of *recombinations*, as with other basic genetic processes (replication, mutation) is that purely structural rearrangements are produced by them; the semantics, the meaning of the genetic messages does not participate in the exchange process at all. The semantic aspect of genetic messages consists of the functions they perform in the genetic system, as opposed to the arrangements in which they are found.

In living organisms a significant excess concentration is characteristic for many proteins. Very frequently for a protein to carry out its function it is sufficient to have only a tiny fraction of its normal concentration. Hence, even if only one of the homologous genes of a zygote is normal, the activity of the protein coded by it is frequently sufficient for normal development. In such cases it is customary to say that the normal gene is *dominant* to the mutant form, which is said to be *recessive*. For some other proteins (for example those which play a regulatory role) the reverse is true: a mutant gene may be dominant to the normal gene, since even a small change of concentration of the normal protein may materially influence the developmental process.

We note however that *dominance* and *recessiveness* are somewhat conditional concepts. They depend on the level of phenotypic expression used in comparing normal and mutant genes. In genetics there are examples of mutations dominant at the molecular level, but recessive from higher-level points of veiw (cellular, tissue, organismal) (see, for example, [30,31]). However, if a standard level of phenotypic expression is chosen then dominance relations will have a stable meaning.

Suppose that in a diploid cell each member of some pair of homologous chromosomes carries one recessive mutation. If the mutations affect corresponding genes the phenotype of the cell will be mutant. If, however, the mutations are located in different genes, then each of the mutated chromosomes will generate an active protein product which the other is not able to generate. As a result, homologous chromosomes act in a mutually complementary way, causing a normal phenotype. Such a phenomenon is called *complementation*.

Thus, in diploids the estimation of the meaning of a genetic message depends not only on the synthesized protein, as such, but also on the confrontation with the product of the homologous gene. The result of this confrontation is manifest phenomenologically as dominance and complementation. Recessive mutations constitute part of the reserve variability of a population, which may turn out to be of vital importance under changes in the external environment.

"Thus we see that the separation of the life cycle into two phases (haploid and diploid) allows, in a remarkable way, the combining of the flexibility necessary for evolution (in sexual cells) with the stability necessary for the individual (and for control of evolutionary changes) . . . that is, the maximal reliability of transmission of hereditary information" [51].

These properties are manifest in particular in the phenomena of recombination, dominance and complementation.

3. Genetic methods of investigation. At present the investigation of genetic systems is conducted by mutually complementary "direct" physico-chemical means and by the "indirect" hybridization methods of genetics.

Genetic methods are based on comparative analysis of the phenotypes caused by mutant genes either singly or in combination. In genetic experiments, normal genes are never studied directly, rather only their "breakages" -- mutations. Comparing these mutations according to their phenotypic effects in various hybrids, the geneticist tries to form a picture of the genetic system being studied, by means of imcomplete, "negative" details.

To understand the essence of the genetic approach one must become somewhat acquainted with the associated terminology.

Earlier we agreed to use the term "gene" to denote a unit of heredity responsible for the manifestation of some single trait. Mutations change the structure of a gene, frequently causing a change in, or complete loss of, the function which that gene controls, thus giving rise to alternative states of that trait. The genes at a given locus associated with such alternative states (for example, eye color in *Drosophila*) are termed *allelic* or (simply) *alleles*. A diploid individual with differing alleles for a single gene is *heterozygous* for that gene, and one with identical alleles is *homozygous*. Alleles, which are situated in identical positions (loci) of homologous chromosomes, are termed *homologous* genes.

Thus, a mutation plays the role of a distinctive marker of the gene (locus) in which it arises. It is these marked fragments of the genetic apparatus which geneticists study, primarily by utilizing the appearance of structural exchange -- recombinations and functional interactions -- in the phenomena of dominance, complementation, etc. Consequently, in genetics there are two basic types of tests: for recombination and complementation.

In a *recombination test* we usually study the results of crossing individuals which differ by two or more mutations. In the descendants, besides the original parent forms, there may appear new ones, *recombin - ant* with respect to the observed trait. The end product of the test is the fraction of recombinant offspring.

The more closely two mutations are situated on a chromosome the less chance there is of a crossing-over occurring on the segment between them, and consequently, the less frequently recombinations will occur. In other words, the frequency of recombination reflects the real physical distance between the positions of the mutations within the chromosome.

It is on this basis that *genetic* (*recombination*) *maps* of chromosomes are built. The recombination test is used to detetermine the frequency of recombination for many pairs of mutations. The collection of all results is then used in building the genetic map. The frequencies in this collection are used to produce the recombination matrix. From this matrix a map is formed -- a segment of the real axis with points (corresponding to separate mutations) drawn on it in such a way that distances between points correspond to frequencies of recombination. It is known that the more complete the initial matrix is the more exact the map will be. However, in practice only a fraction of the possible pairs of mutations are tested.

The construction of the recombination map is based on two properties of the organization of genetic material: its linear structure and the monotone dependence of recombination frequency on physical distance. For many specific situations formulas, called *mapping functions*, have been worked out, which allow the estimation (from the frequencies) of the physical distance between the loci carrying the mutations of interest [17,30,41]. By means of these distances it is easy to precisely arrange the chromosome (or some part of it) along the axis, corresponding to the tested mutations. The result of such an ordering is a map of some segment of a chromosome.

Genetic mapping serves as the basic method of clarifying the structural organization of a genome. This does not exahust its possibilities, however. If the products of separate regions of the map (corresponding to genes and gene systems) are known, then the genetic map also helps to establish a complete representation of the functional organization of the genetic apparatus as a whole (in connection with this, however, see Section 2.3.2). This is why the establishing of suitable recombination maps is the most important step in solving theoretical and practical problems in the fundamental areas of modern genetics. Molecular genetics, classical genetics, the particular genetics of specific species, medical genetics, genetic engineering -- progress in all of these areas depends on the procedure of genetic mapping.

The specific dependence of recombination frequency on physical distance varies greatly among species, among the chromosomes of members of a given species, and even among different sections of the same chromosome. This significantly complicates the task of constructing mapping functions to translate the measured quantities -- recombination frequencies -- into physical distances. The criterion of adequacy for such functions is evident: If the function is correct then the corresponding distances must be additive for all values of the recombination frequencies.

The theory of recombination mapping is quite well worked out for a wide class of genetic objects, and a number of articles and monographs have been written on the subject. We have not taken as our task a detailed treatment of this area, although it plays a central role in the analysis of gene structures. Besides, the "bottleneck" in such mapping tasks -- the construction of mapping functions -- has no direct connection with graph theory. Possibilities are clearly visible today for the use of graph theory in recombination analysis to clarify the topology of genetic maps without the calculation of specific distances between the mutations involved.

Evidently the first person to realize this in a sufficiently clear form was S. Benzer. In his papers [3,55] the notions of graphs and hypergraphs of intervals were actually applied as a means of doing preliminary genetic mapping using only qualitative information about the overlapping of mutational defects, without the calculation of specific recombination frequencies. The theory of graphs and hypergraphs of intervals is discussed in Sections 2-4 and its application is given in Section 5.

We point out another task of recombination analysis (where the methods of graph theory play a significant role) which arises when the chromosome locus being examined is small (in comparison with the sizes of cistrons) and the faulty segments are very near one another (they may correspond to neighboring nucleotides). In this situation the quantitative regularities do not hold, and the construction of exact mapping functions becomes a practical impossibility, since inside cistrons there are frequently observable cases of such phenomena as negative interference, polarity of recombination, allele-specificity of recombination, gene conversions, etc., each of which leads to deviations from the additivity of distance which are difficult to control (for more details see [17]). Then only the qualitative regularity remains: that the closer two defective segments are, the smaller will be the recombination frequency between the corresponding mutations. This means that the matrix of recombination frequencies has a special "linear" structure: From its diagonal the values of elements monotonically increase on "both sides", if the ordering of the mutations corresponds to their true positions in the locus (see also Section 4.3).

The second classical genetic test is the *complementation test* (also called the *cis-trans test*). In this kind of test the fundamental events occur after the formation of zygotes from gametes, each of which carries a different recessive mutation. Parents, mutant with respect to the trait of interest, are crossed. Since each is mutant for a recessive mutation, it must be homozygous at that locus: the corresponding homologous chromosomes carry identical mutant alleles. The offspring of such a cross are hybrids, i.e. heterozygotes whose homologous chromosomes carry different mutations -- one from each parent. If the heterozygote exhibits a normal phenotype then it is evident the mutations affect different genes (that is, they are non-allelic, or complementary). In the other case (non-complementation) the mutations are allelic: they affect the same gene.

The result of pair-wise complementation testing of a collection of mutations is the *complementation matrix*, the (i,j)th element of which designates whether mutations i and j are complementary or not.

If the examined mutations are *point mutations*, i.e. they cause replacement of the same amino acid in the associated polypeptide, then clearly the mutational defect does not extend beyond the bounds of a particular cistron. In that case the complementation matrix must have a particularly simple, block structure: All the mutations affecting a given cistron are pair-wise noncomplementary among themselves and, at the same time, complementary to mutations of other cistrons. Consequently, the set of mutations is arranged into groups corresponding to separate cistrons and giving rise to a block-diagonal form of the complementation matrix. Here, the question of the mutual positions of the cistrons remains open.

The real picture is much more complicated [35,43]. First, the mutations within a given cistron may be complementary. There is a whole class of cistrons in which *intracistronic*, or *interallelic* complementations have been observed. The next chapter is devoted to the possible mechanisms of interallelic complementation and to questions of its analysis by graph-theoretic methods.

In this chapter we examine a second type of deviation from a block-structured complementation matrix, connected with the existence of mutations which affect several cistrons at once. Such mutations "join" separate cistron groups (of matrix elements) and significantly complicate the structure of the complementation matrix used to construct the interallelic topography map of the genetic material.

In this case geneticists are governed by the following rules. Each mutation is characterized by a continuous interval (the defect zone) of the locus under consideration. In the complementation matrix, for each pair of mutations it can be determined whether or not their associated intervals intersect: For noncomplementarity of recessives, in point of fact, means an overlapping of the corresponding defect zones. *Complementation mapping* hinges on the construction, by means of the complementation matrix, of a representation of the given collection of mutations by a system of mutually overlapping intervals.

Although the nature of the complementation test is functional (semantic), this system of intervals provides a portrait of the structural organization of a locus. This is explained by the fact that in intercistronic complementation each cistron acts as a unit, and in that situation the complementation phenomena are not directly connected with the semantics of the cistrons, which are manifest on the protein level.

Therefore, with the degree of reduction to separate cistrons, the complementation map provides the same "structural" information as the recombination map (in other words, both these maps are collinear). This fact is widely used in the practice of genetic analysis, inasmuch as complementation testing is, as a whole, much simpler than recombination testing.

§2. The mathematical theory of linear maps: Interval graphs

1. Maps and interval orders. We consider a set of N objects, given by the indices $1, 2, 3, \ldots, N$. We will designate the set by $A = \{1, 2, \ldots, N\}$. With each $i \in A$ we associate an open interval of the real line I_i, with bounds at points corresponding to natural numbers, i.e. $I_i = (l_i, r_i)$ $(l_i, r_i \in \{0, 1, 2, \quad\})$, where l_i is the left boundary and r_i the right boundary of the interval I_i $(i \in A)$. We designate by $\mathbf{I}$ the minimal interval containing all of the I_i. Then $\mathbf{I}$, together with the intervals I_i $(i \in A)$ forms, by definition, a *map* $K = \langle \mathbf{I}, I_i\ (i \in A)\rangle$. The length of the interval $\mathbf{I}$ will be called the *length of the map K*, so that if $\mathbf{I} = (l, r)$, then the length of the map is equal to $r - l$.

The interval $\mathbf{I}$ is considered to represent a fragment of genetic material, the intervals I_i the defect zones of individual mutations within the bounds of this fragment, and the objects i the numbers (designators) of the corresponding mutations. Keeping this in mind, we will in the future call these objects mutations, not using the term in its special genetic interpretation, but only to aid in the interpretation of subsequent mathematical constructions.

The following basic postulates are associated with the notion of a map (although they frequently may not hold): the linearity of genetic material and the indissolubility of mutational defects.

As a result of genetic investigations of N mutations of some locus (as a rule, from tests of complementation) we frequently obtain an $N \times N$ Boolean matrix $r = \|r_{ij}\|$ (the *complementation graph*), which characterizes the overlapping of mutational distortions: $r_{ij} = 0$ if the ith and jth mutations are complementary (non-overlapping), and $r_{ij} = 1$, otherwise. Correspondingly we say that a map K *represents* a Boolean matrix if and only if

$$r_{ij} = 1 \leftrightarrow I_i \cap I_j \neq \phi. \tag{1}$$

As is usual with the introduction of new mathematical ideas, a number of questions arise. Which Boolean matrices (complementation graphs) are represented by maps (the existence problem)? Is a map which represents a given complementation graph uniquely determined? And if not, what relations hold among the maps representing such a graph (the uniqueness problem)? How is a map of a given Boolean matrix constructed? Etc. In this section we examine questions associated with the "internal" characteristics of matrices (graphs, relations), affecting their representation by maps, and in the next two sections we will take up the questions posed above.

For an arbitrary map $K = \langle \mathbf{I}, I_i (i \in A) \rangle$ we will be concerned with the relation "to-the-right-of" on the set A of objects, defined by

$$(i,j) \in P_K \leftrightarrow I_i > I_j, \tag{2}$$

where the relation ">" is the usual numeric > relation applied to the subset of numbers (intervals), i.e. $I_i > I_j$ means that the entire interval I_i lies to the right of the interval I_j. In other words the left end point of interval I_i is not less than the right end point of interval I_j. This relation P_K we will call the *interval order* for the map K.

The question arises, what form must a relation $P \subseteq A \times A$ take in order for there to exist a map K for which P is the interval order, i.e. $P = P_K$.

It turns out that interval orders are completely characterized by the following *condition of quasi-linearity*:

$$P\langle a \rangle \subseteq P\langle b \rangle \quad \text{or} \quad P\langle b \rangle \subseteq P\langle a \rangle. \tag{3}$$

The quasi-linearity condition means that the sets $P\langle a \rangle$ are linearly ordered by the inclusion relation. Let there be among them exactly m different ones: $P\langle a_0 \rangle$, $P\langle a_1 \rangle$, ..., $P\langle a_{m-1} \rangle$. Then for a suitable renumbering $F(a_0, a_1, \ldots, a_{m-1})$,

$$P<a_0>\subset P<a_1>\subset\cdots\subset P<a_{m-1}>, \tag{4}$$

where the inclusion $\subset$ is strict.

To prove the assertion we will first discuss the set-theoretic structure of quasi-linear relations.

Let $P\subset A\times A$ satisfy (3) and (4). Let

$$A_i = \{a \mid P<a> = P<a_i>\} \quad (i=0,1,2,\ldots,m-1). \tag{5}$$

The collection of these sets $\alpha = \{A_0, \ldots, A_{m-1}\}$ clearly forms a partition of the set A. Let

$$B_j = P<a_j> - P<a_{j-1}> \ (j=1,2,\ldots,m-1). \tag{6}$$

Clearly the sets B_j $(j=1,2,\ldots,m-1)$ are nonempty, since in (4) the inclusions are strict. Further, let

$$B_m = A - \bigcup_{j=1}^{m-1} B_j = A - P<a_{m-1}>.$$

If is antireflexive (and interval orders are), then $B_m \neq \phi$, since $a_{m-1} \notin P<a_{m-1}>$, i.e. $a_{m-1} \in B_m$. Consequently $\beta = (B_1, \ldots, B_m)$ is also a partition of the set A. In view of (4) both of these partitions are *strictly ordered*: the ordering of the classes is determined by the ordering of the sets $P<a>$ under inclusion.

Consider the union of all left and right parts of (6), for j going from 1 to i:

$$\bigcup_{j=1}^{i} B_j = (P<a_i> - P<a_{i-1}>) \cup (P<a_{i-1}> - P<a_{i-2}>) \cup \ldots = P<a_i> - P<a_0>.$$

The set $P<a_0>$ is empty, since if $a \in P<a_0>$ then $P<a>$ $P<a_0>$. Actually, from (3) one of the sets $P<a>$, $P<a_0>$ is contained in the other, but $a \notin P<a>$ because of the antireflexivity of P, so that $P<a_0>$ cannot be contained in $P<a>$. On the other hand the strict inclusion $P<a>\subset P<a_0>$ contradicts the minimality of $P<a_0>$ in accord with (4). Therefore $P<a_0> = \phi$ and for $a_i \in A_i$

$$P<a_i> = \bigcup_{j=1}^{i} B_j \ (i=1,\ldots,m-1). \tag{7}$$

The equalities (7) allow us to express the relation P in terms of the partitions α and β, for from (7)

$$P = \bigcup_{a\in A} \{a\}\times P<a> = \bigcup_{i=0}^{m-1} \bigcup_{a\in A_i} (\{a\}\times P<a>) = \bigcup_{i=0}^{m-1} A_i\times P<a_i>.$$

Hence from (7) we obtain

$$P = \bigcup_{i=1}^{m-1} \left(A_i \times \bigcup_{j=1}^{i} B_j\right). \tag{8}$$

Since P was given as antireflexive the conditions

$$A_i \quad \bigcup_{j=i+1}^{m} B_j \tag{9}$$

hold, for otherwise there must exist $a \in A_i$ such that $a \in \bigcup_{j=1}^{i} B_j$, so that $(a,a) \in P$.

Formulas (8) and (9) completely determine antireflexive quasi-linear relations in the following sense. Suppose we have two ordered partitions of a set A into the same number of classes $R^1 = (R_0^1, \ldots, R_{m-1}^1)$ and $R^2 = (R_1^2, \ldots, R_m^2)$, which satisfy condition (9) (with the replacement of A_i by R_i^1 and B_j by R_j^2). We form the binary relation P, letting, for $a_i \in R_i^1$ $(i=0,\ldots,m-1)$,

$$P\langle a_0\rangle = \phi, \qquad P\langle a_i\rangle = \bigcup_{j=1}^{i} R_j^2.$$

This relation is clearly quasi-linear (by the definition) and antireflexive by (9). Moreover the partitions α and β derivable from it are identical, respectively, to R^1 and R^2.

We have proved the following

<u>Theorem 1</u>. The antireflexive relation P is quasi-linear if and only if there exist uniquely-determined ordered partitions α and β of the set A which satisfy conditions (8) and (9).

Knowing now the structure of quasi-linear relations it is not difficult to obtain the characteristics of interval orders.

<u>Theorem 2</u>. An antireflexive relation is an interval order if and only if it is quasi-linear.

<u>Proof</u>. It is clear that for any map K the interval relation $I_a > I_b$ is antireflexive. On the other hand, corresponding to the set $P_K\langle a\rangle$ is the set of all objects b whose intervals I_b are located to the left of the interval I_a. In view of the one-dimensional nature of the real line such sets are ordered by inclusion, and consequently, the interval order P_K is quasi-linear.

We will now prove the reverse statement: If an antireflexive relation P is quasi-linear then it is represented by the relation ">" on a set of open intervals of the real line. Let a be any element of A. Since α and β are partitions, we can find unique numbers i and j such that $a \in A_i$ and $a \in B_j$. From (9) it is clear that $i<j$. Therefore the correct definition of the interval $I(a)$ is $I(a)=(i,j)$.

We will show that

$$(a,b) \in P \longleftrightarrow I(a) > I(b).$$

Let $(a,b) \in P$ and $a \in A_i$. Then $b \in P\langle a\rangle = \bigcup_{j=1}^{i} B_j$ from formula (8). This means that $b \in B_k$ for $k \leqslant i$. The right end point k of the interval

$I(b)$ is located to the left of the left end point i of the interval $I(a)$, i.e. $I(a)>I(b)$.

Suppose now that $I(a)>I(b)$. Then the left end i of $I(a)$ is to the right of the right end k of $I(b)$: $i \geqslant k$. By definition $a \varepsilon A_i$ and $b \varepsilon B_k$. Since $k<i$, $B_k \subseteq \bigcup_{j=1}^{i} B_j = P<a>$, so that $b \varepsilon P<a>$, i.e. $(a,b) \varepsilon P$. The theorem is proved.

In the proof we constructed a map with intervals $I(a)$ $(a \varepsilon A)$ for the quasi-linear relation P, for which P was an interval order. This map gave meaning to the ordered partitions α and β connected with P. The partition α characterizes the left ends, and β the right ends, of the intervals of the graph.

This map also allows the determination of characteristics of the inverse relation P^{-1} for an antireflexive quasi-linear relation, since it corresponds to the relation "to the left of" for intervals of the map (or to the relation "to the right of" for the map when turned 180°). The relation P^{-1} is also quasi-linear and is determined by the same partitions α and β (with a reversal in their order and the order of their classes).

To conclude this section we note that there is still another characteristic of quasi-linear relations, which is distinct from (3): in "local" terms of objects (and not in "global" terms of sets $P<a>$):

Theorem 3. The relation P is quasi-linear if and only if it satisfies the condition

$$aPb \text{ and } cPd \longrightarrow aPd \text{ or } cPb. \tag{10}$$

Proof. Suppose that P is quasi-linear and aPb, cPd, i.e. $b \varepsilon P<a>$ and $d \varepsilon P<c>$. We will show that aPd or cPb. Suppose the contrary: $(a,d) \notin P$ and $(c,b) \notin P$. This means that $d \notin P<a>$ and $b \notin P<c>$. On the other hand, by assumption, $d \varepsilon P<c>$, so that $P<c> \nsubseteq P<a>$, and $b \varepsilon P<a>$, so that $P<a> \nsubseteq P<c>$, but this contradicts the quasi-linear condition for $P<a>$ and $P<c>$.

Suppose now that for arbitrary a,b,c,d (10) is satisfied. We will show that P is quasi-linear. Suppose that $P<a> \nsubseteq P<c>$, where $P<a> \neq \phi$, $P<c> \neq \phi$, i.e. we can find $b \varepsilon P<a>$ such that $b \notin P<c>$, i.e. $(c,b) \notin P$. Then for any $d \varepsilon P<c>$ from (10) the relation aPd holds, i.e. $d \varepsilon P<a>$, so that $P<c> \subseteq P<a>$, as was to be shown.

We note that for algorithmic testing of the quasi-linearity of a relation condition (3) is more useful than (10) since it requires the fewest examinations of objects and their relationships. The properties of interval orders in terms of (10) were given by P. Fishburn [60] independently of the work of B. G. Mirkin [19], from whom the contents of this section were borrowed.

2. The description of interval graphs. We will examine, for an arbitrary map $K = \langle \mathbf{I}; I_i (i \in A) \rangle$, the intersection relation I_K: $(i,j) \in I_K \leftrightarrow I_i \cap I_j \neq \phi$, defined on the set A of mutations. Clearly I_K is connected with P_K in the following way: $I_K = \overline{P_K \cup P_K^{-1}}$. In other words those and only those intervals intersect, no one of which lies "to the left of" another. The relation I_K is called the *interval equivalence* relation for the map K. It is clear that I_K is reflexive and symmetric. However, it is not necessarily transitive: two intervals do not necessarily intersect though they both intersect a third interval lying "between" them.

What characteristics must an arbitrary reflexive, symmetric relation $I \subseteq A \times A$ have in order for it to be the interval equivalence for some map K? The answer to this question is the more interesting, since the initial complementation matrix gives precisely the intersection relation for intervals of the desired map. In other words, complementation mapping consists in constructing, for a given $I \subseteq A \times A$, a map K for which $I = I_K$.

It is customary to call an ordinary graph corresponding to an interval equivalence an *interval graph*. Following our practice of identifying corresponding notions in the terminology of relations, graphs and matrices (see the appendix), we will consider the notions of an interval equivalence and an interval graph synonomous.

The matrices of interval graphs possess a regular structure. If we renumber the objects (mutations) $i \in A$ corresponding to the ordering of the left ends of their associated intervals, then in the intersection matrix the ones in each row will follow in succession, beginning at the main diagonal:

$$r_{ij} \geqslant r_{i,j+1} \qquad (j=i,i+1,\dots,N-1), \tag{11}$$

for if I_i intersects I_{j+1} for $j \geqslant i$ then because of the renumbering I_i must also intersect I_j, the left end of which lies between those of I_i and I_{j+1}.

Matrices satisfying condition (11) are called *quasi-diagonal*.

Theorem 4. For a complementation matrix to admit a mapping it is necessary and sufficient that it be quasi-diagonalizable under a suitable renumbering of its objects (i.e. simultaneous permutation of rows and columns).

Proof. If a map exists then, as shown above, renumbering the objects to correspond with the ordering of left ends of intervals leads

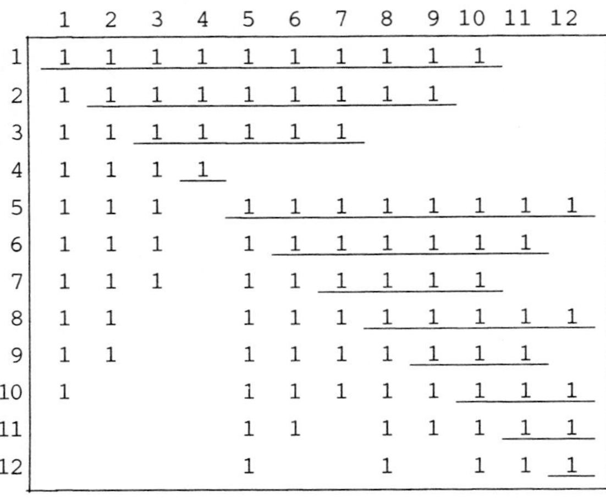

	1	2	3	4	5	6	7	8	9	10	11	12
1	1	1	1	1	1	1	1	1	1	1		
2	1	1	1	1	1	1	1	1	1			
3	1	1	1	1	1	1	1					
4	1	1	1	1								
5	1	1	1		1	1	1	1	1	1	1	1
6	1	1	1		1	1	1	1	1	1	1	
7	1	1	1		1	1	1	1	1	1		
8	1	1			1	1	1	1	1	1	1	1
9	1	1			1	1	1	1	1	1	1	
10	1				1	1	1	1	1	1	1	1
11					1	1		1	1	1	1	1
12					1			1		1	1	1

a)

1

2

3

5

6

7

8

9

10

11

4 12

Complons

b)

Figure 2. The matrix (a) and minimal map (b) of complementation at the locus pan-2 in the fungus *Neurospora crassa* [43]; the matrix is given in quasi-diagonal form.

to a quasi-diagonal matrix.

The converse remains to be proved: If a matrix can be brought to quasi-diagonal form, then there exists a map. To prove this we will show that the quasi-diagonal form of the matrix of (11) itself defines a map. As the intervals of this map we take the segments defined by $I(i)=(i,n_i)$, where n_i is the number (index) of the first zero element to the right of the diagonal in row i, so that $r_{i,n_i-1} = 1$, but $r_{i,n_i} = 0$ $(n_i>i)$. Now the statement that $I(i) \cap I(j) \neq \phi$ means that there exists k such that $i<k<n_i$, $j<k<n_j$. Suppose for example, that $i<j$. Then clearly $i<j<k<n_i$ so that $r_{ij} = 1$ (the mutations i,j are non-complementary).

If, on the other hand, for $i<j$ $r_{ij} = 1$, then $i<j<n_i$ so that (i,n_i) and (j,n_j) intersect. This completes the proof.

The proof of theorem 4 shows that the quasi-diagonal form of a complementation matrix can be easily carried over to a map: It is sufficient, for example, to simply underline, in each row, the set of ones beginning at the diagonal and going to the right (Fig. 2,a). The segments thus defined correspond to intervals.

Hence the mapping problem

is directly connected with the problem of bringing a complementation matrix into quasi-diagonal form, the solution of either problem leading immediately to the solution of the other. However in terms of the complementation matrix itself there is little one can say about the construction of a "quasi-diagonal" ordering of the mutations. In order to describe the processes of local permutations the more "local" language of graphs and relations is required, to which we now turn.

From the fact that $I_K = \overline{P_K \cup P_K^{-1}}$ it is clear that I is an interval equivalence when and only when one can find an interval order (quasi-linear antireflexive relation) P such that $I = \overline{P \cup P^{-1}}$ (see Section 3). Therefore, to investigate I we must move to the complementary relation $\overline{I}$. If I is an interval equivalence, then we can find an interval order P such that $I = \overline{P \cup P^{-1}}$. The graph of P is obtained from the complementary graph $\overline{I}$ by orienting all of its edges. If for a given ordinary graph I there exists such an orientation of its complement $\overline{I}$ satisfying the quasi-linear condition, then I is an interval graph. We will reformulate the condition for the existence of a quasi-linear orientation in the local terminology of condition (10).

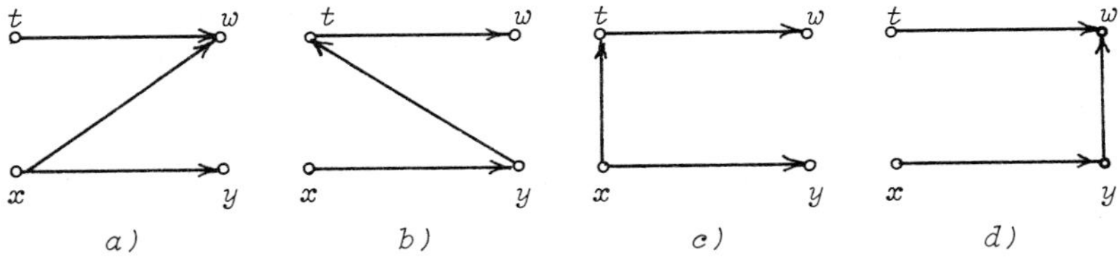

Figure 3

In (10) the objects a,b,c and d are arbitrary and some of them may coincide (though of course $a \neq b$ and $c \neq d$, since P is antireflexive). If, of these four, only two actually differ then clearly condition (10) places no additional restrictions on P. If three of the objects are different it is easy to show by direct substitution that condition (10) becomes simply the transitivity condition for P. For example, if $b=c$, then we have aPb and $bPd \rightarrow aPd$ or bPb, since bPb is impossible aPb and $bPd \rightarrow aPd$, which is transitivity. Finally, if all four of the objects are distinct, condition (10) (together with the transitivity of P) asserts that if in $\overline{I}$ there are two non-adjacent edges on the set $\{a,b,c,d\}$, then there must be a third, adjacent to both of them (i.e. "parallel" edges are always connected by a third edge). If (10) holds, then the presence of a connection between parallel edges in $\overline{I}$ follows trivially from it, if the arcs of P are deoriented. Conversely, if two edges in

$\overline{I}$ are connected, then under orientation of $\overline{I}$ we obtain one of the four configurations of Fig. 3, where the "parallel" edges are shown horizontally, and cases a) - d) reflect all possible placements of the connecting arc, obtained from an original edge by orientation.

In case a) condition (10) is fulfilled automatically; in case c) for vertices x,t,w, in case d) for vertices x,y,w, but in case b) for x,y,t and then x,t,w the relation P must contain, by transitivity, an arc (x,w), leading to fulfillment of (10) for all four objects. Analyzing this argument, we conclude that the existence of a quasi-linear P is equivalent to the existence of a transitive P and the presence of connections between parallel edges in $\overline{I}$ (or, what amounts to the same thing, fulfillment of (10) for any four distinct objects a,b,c,d).

The second condition can be easily reformulated as follows. We will say that an edge u is a *triangulator* of adjacent edges xy and yz if $u=xz$, i.e. the vertices x,y and z form a "triangle." A *triangulator of a cycle* is a triangulator of any two of its adjacent edges.

The existence of a connection between every pair of non-adjacent edges in $\overline{I}$ is equivalent to the existence of a triangulator for every "quadrilateral" (cycle on four vertices) of the initial graph I. The fact that the edges xy, yz, zw and wx form a cycle without a triangulator in I means that in $\overline{I}$ there are two non-adjacent edges xz and yw without any connector, and this proves the assertion. We sum up our discussion in

Theorem 5. An ordinary graph I is an interval graph if and only if it satisfies one of the equivalent conditions:

a) The complementary graph $\overline{I}$ is transitively oriented, with every quadrilateral in I containing a triangulator;

b) The complementary graph $\overline{I}$ is transitvely oriented, with each of its transitive orientations P satisfying the quasi-linear conditions (10) or (3).

Condition b) needs some comment. The initial assertion of the theorem requires the existence of at least one quasi-linear orientation. Why then did b) refer to the quasi-linearity of every result of a transitive orientation of $\overline{I}$? The fact is that the triangularity condition for quadrilaterals in I is in no way connected with the orientation of $\overline{I}$, and therefore, in connection with the discussion of Fig. 3 we must conclude the quasi-linearity of any transitive P for which $P \cup P^{-1} = \overline{I}$.

This type of result: from one fact formulated in terms of the initial relation I, are deduced consequences true for an arbitrary orientation of $\overline{I}$, will be encountered repeatedly in the following analysis.

We remark that condition a) was first proved by Gilmore and Hoffman [67]in a direct investigation of interval graphs (but not by reducing it to a case of quasi-linearity).

Theorem 5,a) shows that the checking of the interval equivalence of I leads to the investigation of all quadrilaterals in I, and to the elucidation of whether $\overline{I}$ is transitively orientable.

As will become clear in what follows, the most simple transitive orientability is verified by direct construction of the corresponding transitive P. In this case it is not necessary to consider all possible quadruplets of objects: In accord with condition b) of the theorem it is sufficient to verify the quasi-linearity of this P. This is much easier than examining quadrilaterals, since it only requires a single examination of P, with the goal of constructing all sets of the form $P\langle a\rangle$, and the subsequent verification of whether or not they are completely ordered by inclusion. The construction of the sets $P\langle a\rangle$ also directly provides a map, representing I, as was described in the proof of theorem 2.

However, the formulation a) of theorem 5 represents supplementary interest as a step toward characterizing interval graphs using only interval graph terminology, without bringing to bear "additional" operations such as going to complementary graphs and complete or partial orderings of vertices. For such an "internal" description in agreement with this formulation it remains to characterize those graphs which admit transitive orientations. This question has independent significance of its own, but the extent of the present work precludes its examination. Existence criteria and algorithms for transitive orientation are considered in [7]. We will use the *Gilmore-Hoffman criterion*, which states that an ordinary graph admits a transitive orientation if and only if every nonrepeating cyclic route of odd length in it has at least one triangulator [67,7].

At present we know that I is an interval graph if and only if all quadrilaterals in I and all cyclic routes of odd length in $\overline{I}$ admit triangulators. The remaining step is to move from conditions on $\overline{I}$ to conditions on I.

The conditions for the transitive orientability of $\overline{I}$ mean that for any sequence of vertices $a_0, a_1, \ldots, a_{2n}$ such that among the pairs (a_i, a_{i+1}) $(i=0,\ldots,2n)$ none are identical, the condition

$$(a_0, a_1) \in \overline{I} \text{ and } (a_1, a_2) \in \overline{I} \text{ and } \ldots \text{ and } (a_{2n}, a_0) \in \overline{I} \longrightarrow$$
$$\exists i \leqslant 2n\left[(a_i, a_{i+2}) \in \overline{I}\right]$$

is fulfilled. For indices k greater than $2n$ we will find it necessary to use instead indices obtained as remainders on division of $k-1$ by $2n$.

For example, *0* in place of *2n+1*, *1* in place of *2n+2*, etc.

In terms of I this condition, having the form "u→v", is reformulated in the inverted form of "not v→not u":

$$\left(\forall i\ [a_i I a_{i+2}]\right) \rightarrow \exists j\ [a_j I a_{j+1}].$$

We renumber the elements of the sequence $a_0, \ldots, a_{2n}$ as follows: At the beginning we place elements with even indices, and following them, those with odd indices (in order by indices): $a_0, a_2, a_4, \ldots, a_{2n}, a_1, \ldots, a_{2n-1}$. Then the condition of transitive orientability of $\bar{I}$ in the terminology of I means that if the sequence $a_0, a_2, \ldots, a_{2n}, a_1, \ldots, a_{2n-1}$ is such that among the pairs (a_i, a_{i+1}) there are no repetitions, and it forms a cyclic route, then in I we can find an edge which divides it "into two parts" (the edge $a_j a_{j+1}$ joins an "even" vertex with an "odd" vertex and divides the set of vertices of the route into two parts with n and $n+1$ vertices).

It is easy to conclude that each of the parts thus obtained also gives a cyclic route, and one of them contains an odd number of vertices, so that there must again exist an edge dividing it in two, and so forth, to the point when one of the parts contains three vertices, so that the dividing edge (which led to this part) is also a triangulator of the initial route. We thus arrive at the result that fulfillment of the Gilmore-Hoffman criterion for $\bar{I}$ leads to its fulfillment for I also, so that I and $\bar{I}$ simultaneously admit (or do not admit) a transitive orientation.

However this is not so. Examine the graph of Fig. 4 a), which represents the map of Fig. 4 b).

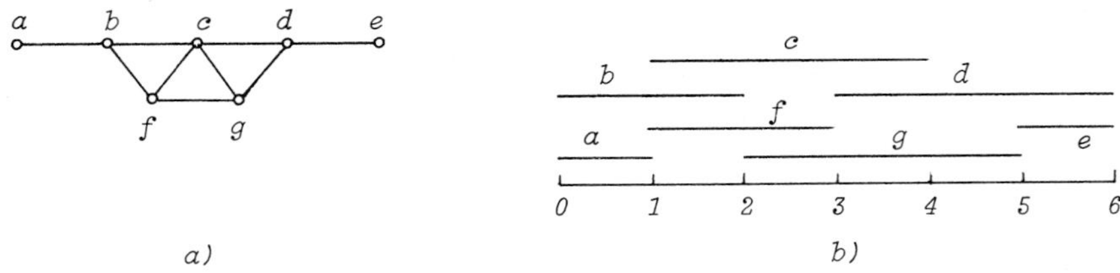

Figure 4

In this graph the cyclic route *abcdedgfba* has no triangulator, although its complementary graph unquestionably admits a transitive orientation (the ordering *abfcgde* leads to quasi-diagonal form for the matrix of the graph).

The mistake in our earlier argument lies in the assertion that the process of successive divisions necessarily leads to a triangle. This is true only in the case when each dividing edge is not contained in the route, as for example, when the route is properly a cycle. If some edges are encountered twice (in a nonrepeating route this is the maximum number of times an edge can be encountered), then the dividing edge may belong to the route and yet not be a triangulator. In our example either of the edges *fc* and *cg* are dividers for the route *abcdedgfba*. Selecting *fc*, we obtain *abcfba* as a route of odd length, for which the edge *bf* is already a divider. Yet *bf* belongs to the route and is not a triangulator.

A similar mistake appeared in [83] and was corrected by Fisher [61]. Thus, in general, the conclusion we reached does not hold. However, it is true for cycles of odd length: if the Gilmore-Hoffman condition is satisfied for $\overline{I}$, then in I each cycle of odd length has a triangulator. In fact there is a stronger property of graphs of intervals: each cycle in I has a triangulator.

To prove this assertion we suppose there exists in I a cycle of odd length without a triangulator (it is also possible to prove this by direct consideration of the map representing I). Consider the one (cycle) among them having minimal length. This cycle clearly has no chords. For, from its minimality, each of its chords must divide it into two "subcycles" of odd length, and these cycles, from the foregoing, must have triangulators. These triangulators are either triangulators of the initial cycle, which is not possible by definition, or they are chords of the initial cycle, which divide it into subcycles of even length, which is impossible due to its minimality.

Thus, in $\overline{I}$ all the vertices of a given cycle, except neighboring ones, are connected with one another by edges. We consider the first five vertices *1,2,3,4,5* (the situation of only four vertices is not possible by theorem 5,a). From the above, there is in $\overline{I}$ a cyclic route of length 5: *352415* which has no triangulators: any of its pairs which are connected by edges have, as triangulators, edges of the original cycle in I. This contradicts the Gilmore-Hoffman condition.

Thus, in an interval graph all cycles must have triangulators. A graph in which all cycles possess triangulators is said to be *triangulated* We have shown the triangulatedness of interval graphs.

Is the triangulatedness condition for a graph not only necessary but also sufficient to characterize interval graphs?

To answer this question we have to determine whether, in the complement $\overline{I}$ of a triangulated graph I, the cyclic routes of odd length must be triangulated or not.

An edge of a cyclic route in $\overline{I}$ either is encountered twice, or belongs to a cycle within the route, and consequently, if the route has no cycles of odd length, the overall length of the route must be even. Therefore a cyclic route of odd length must contain a cycle of odd length. This cycle clearly must be trangulated, since in the complementary graph $\overline{\overline{I}} = I$ all cycles of odd length are triangulated, and this, as we saw, leads to the existence of triangulators also in the cycles of the "original" graph $\overline{I}$.

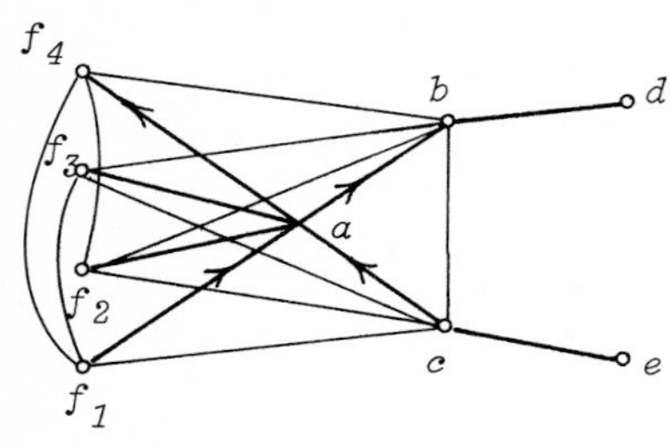

Figure 5

Are the triangulators of this cycle also trinagulators of the route? Not if the triangulated neighboring edges are not encountered in the route one after the other, but "separately." For this to be possible it is necessary to have edges in the route which are incident on the cycle at vertices joining these edges, as illustrated in Fig. 5.

Here ca and ab are successive edges of the cycle under consideration, with triangulator bc. The route "enters" the cycle along edge f_1a, continues in the "upper" part of the cycle $ab\ldots$ (possibly with exits from the cycle such as $af_4af_3af_2af_1$ at the point a, because of the necessity of the absence of triangulators), and returns to f_1 by the route $caf_4af_3af_2af_1$. To make Fig. 5 more readable edges belonging to the route are drawn darkly.

In the simplest case a cycle traversable by a route is a triangle abc such that for all three vertices of the cycle there are "branches" of the type illustrated for vertex a of Fig. 5 (naturally, the number of edges leaving these vertices need not be four) in order to have the non-triangulatedness of the route.

It is easy to understand that in the language of the original graph I, such a route in $\overline{I}$ is non-triangulated if and only if the vertices a,b,c form in I a so-called *asteroidal triplet*. A collection of three pair-wise nonadjacent vertices a_1, a_2, a_3 is called an asteroidal triplet of a given graph, if for each pair a_i, a_j $(i \neq j)$ there exists a chain joining them, such that none of its vertices is adjacent to the third vertex a_k $(k \neq i,j)$. For example, for the graph $\overline{I}$ of Fig. 5 the vertices b and c are joined in I by the chain $bf_1f_2f_3f_4c$ not adjacent to a.

Thus, for I to be an interval graph it is necessary for it not only to be triangulated but also that it have no asteroidal triplets: This ensures the absence in the complementary graph of non-triangulated routes of odd length encountered in three-vertex cycles. In fact, a more general statement is true: The absence of asteroidal triplets in a triangulated I ensures the absence of non-triangulated routes of odd length in I, i.e. the representability of I by a map. This means that the following criteria, formulated in terms of I, hold.

A graph is said to be *asteroidal* if it contains asteroidal triplets.

Theorem 6. An ordinary graph is an interval graph if and only if it is triangulated and is not asteroidal.

This theorem was first established by Boland. Its proof, based on a direct inductive (on the number of vertices) construction of a map representing a given triangulated non-asteroidal graph, was published in a paper by Lekkerkerker and Boland [80]. A main notion in their construction was that of a simplicial point, corresponding to a complon of the map (see below, §3).

We leave the proof of Theorem 6 from Theorem 5 as an (quite trivial) exercise for the reader (see also [96]).

In concluding this section we note that Theorem 5 allows one to obtain very elegant criteria (although little suited to practical constructions) for the representability of a graph in terms of the absence in it of graphs of a special form (that reminds one of the well-known planarity criterion for graphs [78]).

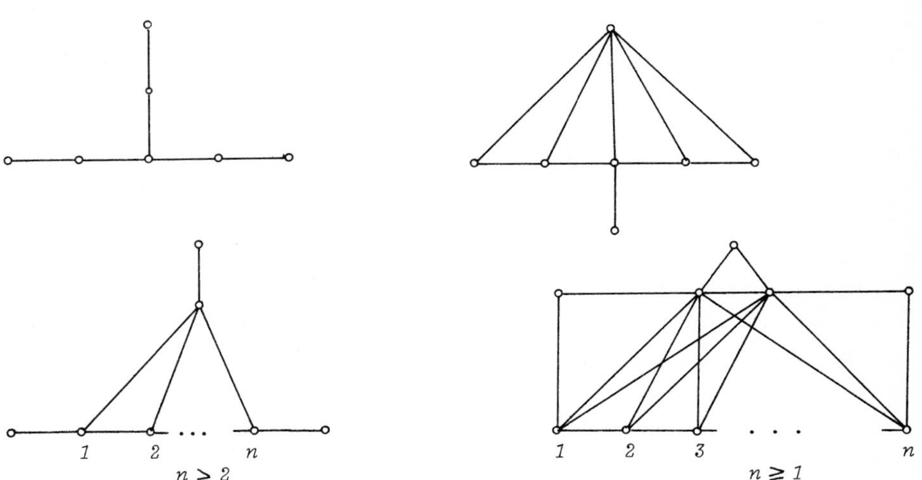

Figure 6

In [80] it was established that triangulated non-asteroidal graphs containing no proper asteroidal subgraphs, cannot contain subgraphs of any of the four types illustrated in Fig. 6. This means that the following criterion holds:

Theorem 7. A graph is an interval graph if and only if it contains no subgraphs of the types illustrated in Figs. 6 and 7.

Fig. 7 illustrates the minimal non-triangulated graphs.

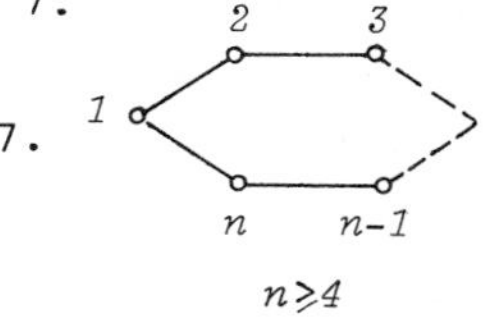

3. Graphs of non-covering intervals. We will consider those interval graphs which are represented by maps with intervals I_i $(i \varepsilon A)$, no one of which is contained in another, i.e. $I_i \subseteq I_j \rightarrow I_i = I_j$. We call these *graphs of non-covering intervals*.

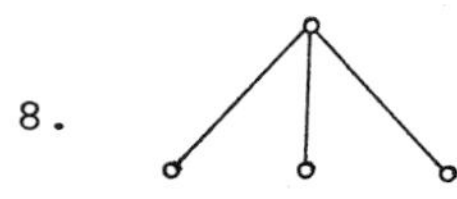

Figures 7 and 8

For such graphs it is not possible, in particular, to have a situation in which one interval intersects three other mutually intersecting intervals, i.e. inside one mutational defect there are not three other mutually complementary distortions. This fact characterizes graphs of non-covering intervals.

Theorem 8. An interval graph I is a graph of non-covering intervals if and only if it contains no subgraphs of the form shown in Fig. 8.

This theorem was proved by Roberts [89] and at the same time in undergraduate work done under the direction of one of the authors by V. A. Kogan (Novosibirsk State University, 1970). It will not be proved here. We note only that the proof by V. A. Kogan was a simple translation to the language of interval equivalences of the following result about interval orders (analogous to obtaining Theorem 5,a from Theorem 4).

Theorem 9. An antireflexive quasi-linear relation $P \subseteq A \times A$ is a relation of an interval order for a map of non-covering intervals if and only if it satisfies one of the following equivalent criteria:

$$(a,b) \ \varepsilon \ P \text{ and } (b,c) \ \varepsilon \ P \longrightarrow (a,d) \ \varepsilon \ P \text{ or } (d,c) \ \varepsilon \ P, \tag{11}$$

$$P\langle a\rangle \subseteq P\langle b\rangle \text{ or } P^{-1}\langle a\rangle \subseteq P^{-1}\langle b\rangle. \tag{12}$$

Relations of an interval order for a map of noo-covering intervals were first considered in the theory of psychological measurements by R. Luce [82], and by P. Suppes and D. Scott [91], who called them semi-orders. The characterization of semi-orders in the "local" terms of (10) and (11) was done by the authors. Conditions (3) and (12) were obtained in B. G. Mirkin [19].

The necessity of condition (12) for non-covering intervals is determined by its geometric meaning: For any intervals a and b the right or left end of interval a is to the right of or to the left of the corresponding end of interval b. The sufficiency follows easily from the structure of Theorem 3: If $I(a) = (i,j)$ and $I(b) = (i',j')$ with $i<i'$,

$j>j'$, then from (8) $P^{-1}\langle a\rangle \subset P^{-1}\langle b\rangle$ and $P\langle a\rangle \subset P\langle b\rangle$, where the inclusion is strict, since for b and a condition (12) is not satisfied.

This shows that on the map obtained in Theorem 3 the covered interval and the covering interval must coincide at their left or right ends. Such a map can be changed into a map for non-covering intervals without violating the intersection relation of the intervals. If, for example, the left ends i, i' of the intervals $I(a)$ and $I(b)$ coincide, where $I(a) \subset I(b)$, then each interval whose left end is not greater than $i=i'$, "behaves" identically with respect to $I(a)$ and $I(b)$: simultaneously intersects or does not intersect them both. Therefore we can introduce at the point $i=i'$ a unit interval, including it in all intervals of the map which contain $i=i'$ as well as in $I(a)$. Now $I(a) \not\subset I(b)$ and the intersection relationship is unchanged.

By carrying out suitable insertions for all cases of inclusion of intervals we obtain a map of the desired form, as was to be shown. We leave the proof of the equivalence of (11) and (12) to the reader.

We will also give, without proof, a characterization of the matrices of non-covering intervals.

A Boolean matrix will be called *row-linear* (or *column-linear*) if in each of its rows (or columns) the units (ones) are arranged consecutively. A matrix is said to be *linear* if it is row-linear or column-linear. For symmetric square matrices the three notions coincide.

<u>Theorem 10</u>. A graph is a graph of non-covering intervals if and only if its matrix can be brought into linear form by simultaneously permuting rows and columns.

Clearly linear matrices are quasi-diagonal. However, the construction of an ordering of vertices (i.e. of rows and columns) which will produce a matrix of linear form is easier than ordering the vertices of an arbitrary interval graph. An algorithm for such a construction is described in detail in Section 4.1.

§3. The mathematical theory of linear maps: interval hypergraphs

1. <u>Maps and interval hypergraphs</u>. The results described in the preceding section give the characteristics of those complementation data representable by maps. They may also be used to construct such maps. These results are not sufficient, however, for the analysis of real complementation data.

This is primarily because of the many possible representations (in general) of a complementation matrix by graphs: A single initial data set may correspond to many maps on which the same mutations are representable by many intervals of various lengths. To obtain an encompassing genetic interpretation one must try to identify the invariants of the maps, determined only by the original complementation matrix.

Of no less interest is the question of what happens if a complementation graph is not an interval graph (possible reasons for this are discussed in Section 5.2). Real complementation matrices, involving 20 or more mutations cannot usually be brought into quasi-diagonal form. Can the techniques of interval graphs really help in the investigation of this situation?

The difficulties we described (non-uniqueness and impossibility of an interval representation of complementation data) are not so heterogeneous as may appear at first glance. The fact is that the invariants of maps must somehow or other reflect the influence of the elementary complementation units, i.e. "*complons*", and consequently admit characterization in terms of "complons." The non-representability of complementation graphs is naturally interpreted in these terms as the impossibility of a non-contradictory "linear arrangement" of complons. This naturally leads to consideration of the possibility of a "nonlinear" arrangement of complons for real complementation matrices.

To pursue this path we must study the maps themselves and not just the intersection graphs of their intervals.

The set-theoretic object corresponding to a map is an interval hypergraph. We will call a system of subsets $S_1, S_2, \ldots, S_N$ of a set $X = \{x_1, x_2, \ldots$ $\ldots, x_n\}$ an *interval hypergraph*, $\Gamma = (X, \{S_i\}_{i \in A})$ if there exists an ordering of the set X of vertices for which the sets $S_1, S_2, \ldots, S_N$ are intervals. As usual a set S is called an *interval of the ordering* P, if for any two objects x, y contained in S, all objects in P lying between x and y are also contained in S.

There is a one-to-one correspondence between maps and interval hypergraphs. The set of vertices of an interval hypergraph Γ is formed by all the unit map intervals x_i of the form $(i, i+1)$ $(i=0, 1, \ldots)$, and a set S_j consists of those unit intervals x_i for which $x_i = (i, i+1) \subset I_j$. Conversely, every hypergraph of intervals gives a map whose intervals I_j correspond to the sets S_j $(j=1, 2, \ldots, N)$.

What features are peculiar to the matrices of interval hypergraphs? Clearly for that ordering of the rows (and of the set X) for which all the S_j become intervals of X, in each column of the matrix the ones are situated consecutively. In the inverse case a set S_j corresponding to a column in which this condition is violated is not an interval for the given ordering of X. This means that a rectangular Boolean matrix is the matrix of an interval hypergraph if and only if it can be brought to column-linear form by some permutation of its rows. This permutation of the rows completely characterizes the map of the interval hypergraph

Thus an interval hypergraph is still not a map: The set X of vertices of the hypergraph differs from the base interval $\mathbf{I}$ of a map in that X is not ordered as $\mathbf{I}$ is. X is an unordered collection of unit intervals.

There are still other properties of interval hypergraphs. We consider the relations γ_X and γ_A associated with the hypergraph $\Gamma = (X, \{S_i\}_{i\varepsilon A})$;

$$(S_k, S_l)\varepsilon\gamma_A \longleftrightarrow S_k \cap S_l \neq \phi,$$

$$(x_i, x_j)\varepsilon\gamma_X \longleftrightarrow \exists k\varepsilon A[x_i, x_j \varepsilon S_k].$$

Clearly Γ is an interval hypergraph if and only if γ_A is an interval graph. Moreover, any map representing γ_A yields an interval hypergraph Γ' with $\gamma_A' = \gamma_A$.

At the same time, the relation γ_X does not generally provide information about the characteristics of the interval hypergraph. By the definition of γ_X

$$\gamma_X = \bigcup_{k\varepsilon A} S_k \times S_k.$$

Thus if $S_l \subseteq S_k$, knowledge of γ_X does not provide any information about S_l, since $S_l \times S_l \subseteq S_k \times S_k$. The relation γ_X is defined by only those "edges" S_k ($k\varepsilon A$) which are not proper subsets of other edges. We will designate the collection of all indices of such maximal edges of the hypergraph by A'. Then

$$\gamma_X = \bigcup_{k\varepsilon A'} S_k \times S_k$$

Moreover, from γ_X it is easy to reconstruct all the maximal edges: They are simply the maximal cliques of the graph of γ_X. Non-maximal edges are subsets of the maximal cliques, but information only about γ_X does not allow the determination of which subsets are edges and which are not.

An ordering of the set X for which all the maximal cliques of γ_X turn out to be intervals, provides a partial interval hypergraph only for the maximal edges. The non-maximal edges will not necessarily be intervals of this ordering.

Ordering the elements k of the set A' according to the order assigned the left ends of the intervals S_k in X, we find that in the matrix of the partial interval hypergraph $\Gamma' = (X, \{S_k\}_{k\varepsilon A'})$ the ones in each row are consecutive, since the S_k do not cover one another, and consequently the same thing holds for an ordering of the right ends of S_k $(k\varepsilon A')$. This means that Γ' is row-linear, and consequently its transpose matrix Γ'^T also corresponds to an interval hypergraph

$$\Gamma'^T = (A', \{T_i\}_{i\varepsilon X}),$$

where

$$T_i = \{k \mid x_i \varepsilon S_k\}.$$

Thus the graph γ_X is a graph of non-covering intervals if and only if the partial hypergraph defined by its maximal edges is an interval hypergraph.

In particular, if all the edges of the original hypergraph are maximal (i.e. the hypergraph corresponds to a graph of non-covering intervals) then the graphs γ_X and γ_A are simultaneously interval graphs or not interval graphs.

The matrices G of the relations γ_X and γ_A may be considered as Boolean products of the $n\times N$ matrix of the hypergraph Γ and the transposed $N\times n$ matrix Γ^T

$$G_X = \Gamma\times\Gamma^T \quad \textit{and} \quad G_A = \Gamma^T\times\Gamma.$$

Boolean matrix multiplication is the same as the usual matrix multiplication except that in place of (scalar) multiplication the operation of taking the minimum (the *conjunction* or ANDing) is used, and in place of addition the operation of taking the maximum (the *disjunction* or Oring) is used. In fact, the (i,j)th element of the Boolean matrix $\Gamma\times\Gamma^T$ is equal to 1 if and only if there exists a column of the matrix Γ in which ones appear at positions i and j, i.e. there exists an edge S_k containing both x_i and x_j. Similarly, the (k,l)th element of the Boolean matrix $\Gamma^T\times\Gamma$ is equal to 1 if and only if there exists a row in which ones appear at positions k and l, i.e. there exists an x_i contained both in S_k and S_l, so that $S_k \cap S_l \neq \phi$.

Curiously, in an analogous way an interval hypergraph Γ can be characterized in terms of the usual matrix products $\Gamma\Gamma^T$ and $\Gamma^T\Gamma$. For an arbitrary hypergraph Γ the (i,j)th element of the $n\times n$ matrix $\Gamma\Gamma^T$ specifies the number of edges of the hypergraph containing both x_i and x_j, and similarly the (k,l)th element of the $N\times N$ matrix $\Gamma^T\Gamma$ specifies the cardinality $|S_k \cap S_l|$.

Clearly if Γ is an interval hypergraph, and X is ordered according to an interval representation of Γ on a map, then the symmetric matrix

$\Gamma\Gamma^T$ has a simple organization: In each row the maximal element appears on the main diagonal, and the values of elements decrease (more accurately, do not increase) monotonically in both directions from this diagonal element. This property is an extension of the linearity property of Boolean matrices. We will therefore call a square Boolean matrix $B = \|b_{ij}\|$ *row-linear* (*column-linear*) if in each of its rows (columns) the elements are monotonically non-increasing in both directions from the diagonal element, i.e.

$$b_{ij} \geqslant b_{ik} \text{ for } i \leqslant j \leqslant k \text{ or } i \geqslant j \geqslant k$$

for rows, and

$$b_{ji} \geqslant b_{ki} \text{ for } i \leqslant j \leqslant k \text{ or } i \geqslant j \geqslant k$$

for columns.

A *linear* matrix is one that is both row-linear and column-linear. For symmetric matrices all three conditions are, of course, equivalent.

We give the linearity property of the matrix $\Gamma\Gamma^T$ for interval hypergraphs in the form of the following theorem [75]:

Theorem 1. If the Boolean matrix Γ can be brought into column-linear form by some permutation of its rows, then the same permutation (of rows and columns, simultaneously) will bring the matrix $\Gamma\Gamma^T$ into linear form.

The analogous property does not generally hold for the matrix $\Gamma^T\Gamma$ because of the possibility of the covering of intervals. For example, the intervals of the ten-element set $X = \{1, \ldots, 10\}$, $S_1 = (1,2,3,4,5)$, $S_2 = (2,3,4)$, $S_3 = (4,5,6,7,8)$, $S_4 = (5,6)$, $S_5 = (6,7,8,9,10)$ form the submatrix of $\Gamma^T\Gamma$:

$$\begin{Vmatrix} 5 & 3 & 2 & 1 & 0 \\ 3 & 3 & 1 & 0 & 0 \\ 2 & 1 & 5 & 2 & 3 \\ 1 & 0 & 2 & 2 & 1 \\ 0 & 0 & 3 & 1 & 5 \end{Vmatrix},$$

which cannot be brought to linear form (by the theory of linear matrices, see §4).

It can be proved, nevertheless, that the matrix $\Gamma^T\Gamma$ carries quite complete information about the hypergraph Γ [66]:

Theorem 2. Suppose, for two rectangular Boolean matrices Γ and Δ, the condition $\Gamma^T\Gamma = \Delta^T\Delta$ is satisfied. Then these two matrices simultaneously either are or are not matrices of interval hypergraphs. If Γ is an interval hypergraph and Δ has the same number of rows as Γ, then Δ is a matrix of the same interval hypergraph as Γ.

The proof follows from the results of §4 (page 59).

A description of interval hypergraphs without regard to the ordering of objects, may be obtained, by using Theorem 2.7, on the basis of the analysis of hypergraphs whose γ_A-graphs are illustrated in Figs. 6 and 7.

A. Tucker proved the following statement [96]:

Theorem 3. A rectangular matrix is the matrix of an interval hypergraph if and only if it contains no submatrices of the forms I_n, II_n, III_n $(n \geq 1)$ and IV, V:

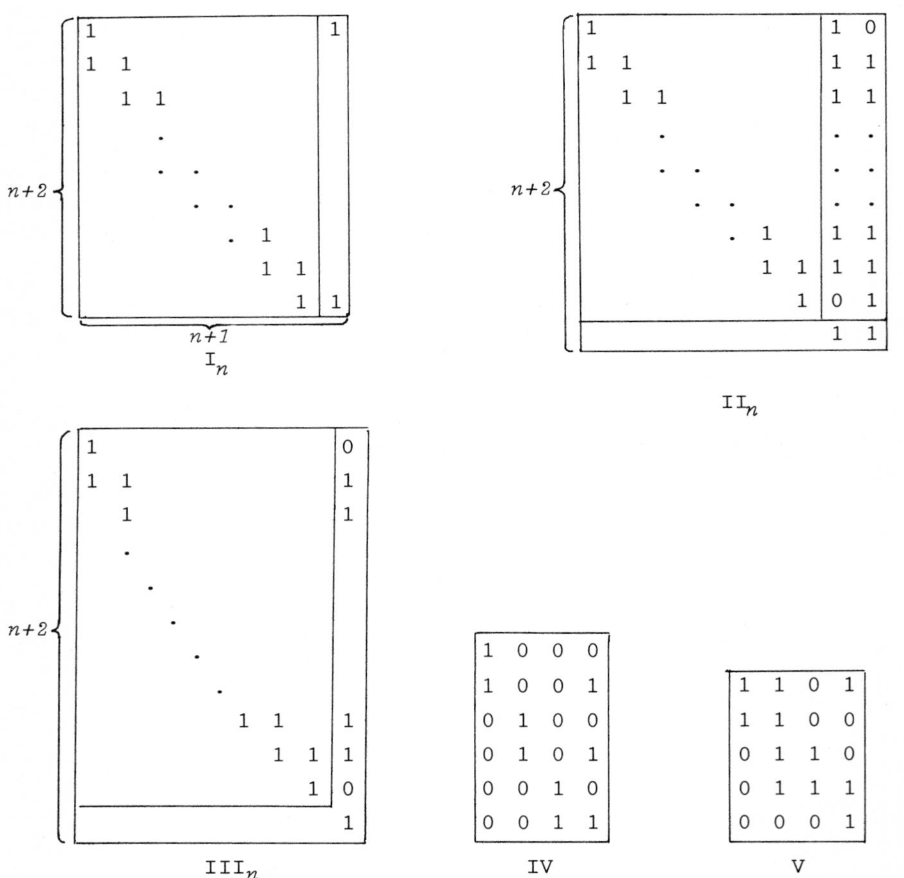

2. Minimal hypergraphs and complons. The invariants of the initial information are not reflected in every representative map, but only in the minimal one. This and the following subsections of this section are

devoted to sharpening this thesis.

For a given interval graph I, we will say that a representative map is *minimal* if it has the shortest length among all the representative maps of I. The set of vertices of the associated hypergraph has minimal cardinality with respect to all Γ's for which $\gamma_A = I$. This interval hypergraph will also be called *minimal*.

It may seem that to obtain the minimal map we need only remove some objects from A, leaving only those whose *neighborhoods* $I\langle a\rangle$ differ; and similarly: some of the vertices of X, leaving only those to which correspond differing rows of the representative hypergraph. But this is not so.

In fact, the construction of the minimal map is given by Theorem 3 §2: for any map, in terms of the associated interval order the left ends of those intervals with identical inverse images $P^{-1}\langle a\rangle$ are identified (as one), and similarly, with the right ends of intervals whose images $P\langle a\rangle$ are identical. By this technique we obtain a map of length m, and this length cannot be decreased without violating the intersection relation. The fact is that, by the construction of this map, for every "internal" natural number k $(0<k<m)$ there exist intervals for which k is the left end and intervals for which k is the right end (since the classes A_k and B_k are non-empty).

However, it does not follow that the map thus constructed is minimal: The size of m may, generally speaking, vary from map to map (the maps being taken as initial maps). The minimality of such a map follows later (see Corollary 2), and therefore we will not give a direct proof here.

For the time being we simply note the mentioned necessary condition for minimality of a map:

<u>Lemma 1</u>. A map $\langle \mathbf{I}, I_i (i\varepsilon A)\rangle$ is non-minimal if there exists an internal natural number $k\varepsilon\mathbf{I}$ which is not the right (or left) end of any of the intervals I_i $(i\varepsilon A)$.

In fact, if k is not the right end of any interval, the intersection of the intervals is still maintained if we eliminate the unit interval $(k-1,k]$, i.e. "shorten" the map so that the points $k-1$ and k fall together (without changing any remaining parts of the map).

Actually, the exclusion of a unit interval can only diminish the number of intersecting pairs of intervals. For intervals I_i and I_j become non-intersecting if the left end of one of them is at $k-1$ and the right end of the other is at k. But this is not possible by definition. Q.E.D.

In reality the reverse assertion is true: A map is minimal if and only if each internal natural k is simultaneously both a left and a right end of some intervals of the map (see Corollary 2).

We will call the unit intervals of a minimal map *complons* [23,40]. They correspond to the vertices of the associated hypergraph.

The genetic equivalent of a complon is an elementary portion of a genetic locus, distinguishable by complementation testing of a given set of mutants. It is known that if the number of mutants is varied the number of complons may vary also.

An object $i \in A$ is said to be *complonic* if the associated interval of the minimal map is a complon (later we will show (Corollary 1) that this definition is justified: A complonic object in any minimal map is represented by a complon).

Let $I \subseteq A \times A$ be a reflexive symmetric relation, characterized by a complementation matrix. Consider the following condition:

$$I\langle i\rangle \subseteq \bigcap_{j \in I\langle i\rangle} I\langle j\rangle \tag{1}$$

To check this condition only the initial matrix is used, not requiring even that it be reducible to quasi-diagonal form.

Clearly, vertices satisfying (1) are characterized by the condition that the collection $I\langle i\rangle$ be a complete subgraph-clique. This clique is clearly maximal. Such vertices were called *simplicial vertices* by Boland and Lekkerkerker [80]. They used the properties of such vertices in proving their interval criteria for graphs.

Independently in [50] it was noted that a *mutation* $i \in A$, for which (1) is not satisfied, cannot be last in the quasi-diagonal form of the complementation matrix. This fact was of fundamental use in [50], where V. V. Shkurba developed an algorithm to bring a matrix to quasi-diagonal form. In accordance with this algorithm, at each step of the construction of the desired ordering the simplicial points (among those points not yet ordered) are examined. Each of these provides its own "branch" for construction of the desired ordering.

The authors observed in [23] that condition (1) actually characterizes complonic mutations.

Theorem 4. An object $i \in A$ is complonic if and only if it satisfies condition (1).

Proof. If an object $i \in A$ is complonic then (1) clearly holds, since every interval of the map intersecting $I_i = (k,k+1)$ contains I_i.

Conversely, let $i \in A$ satisfy condition (1). Suppose it is not complonic. This means that for some k I_i includes the interval $(k,k+2)$ of length 2. By the minimality of the map, and Lemma 1, there exists an interval I_s for which the point $k+1$ is the right end, and an interval I_t for which $k+1$ is the left end.

These intervals do not intersect: $I_s \quad I_t = \phi$; yet clearly $s,t \varepsilon I\langle i \rangle$ and by (1) s and t must be adjacent. This contradiction proves the theorem.

Condition (1) is formulated in terms of the initial matrix of complementations and is not dependent on a specific map. Therefore the following corollary flows directly from Theorem 4:

Corollary 1. Every minimal map which represents a given interval graph has the same set of objects as its complons.

We have already noted that the real prototypes of complons are sections of genes, which are able to mutate both separately (complonic mutations) and in various combinations (non-complonic mutations). Therefore in the general case Corollary 1 shows the possibility of dependably revealing such elementary fragments of the genetic system.

It is necessary, however, to make a significant stipulation. In intracistronic complementation phenotypic effects are connected with the quaternary structure of the proteins, or more precisely, with the constitution of the functional centers of protein molecules (see Section 2.1.1), each of which appears under complementation testing as a unit. Consequently it would seem, for sufficiently complete and detailed initial data, including a large number of mutations affecting only one center, condition (1) would allow the unique identification of all the special centers of a protein molecule. From this one understands the importance of Corollary 1, according to which, such real complons are independent of the specific map invariants.

However, as we shall see below (Section 2.1.1), this very rule of complementation mapping is found to be in serious conflict with mechanisms of mutational complementation at the level of proteins with quaternary structure. Therefore in §2.1 we give a more adequate procedure for representing intracistronic complementation by graphs of another type, which reflect the real picture of the interconnections of the functional centers in complex proteins.

Complons of a minimal map are *real* if they correspond to complonic objects (i.e. the corresponding unit intervals are map intervals). Other (non-real) complons are *fictitious*.

We first examine the case when all the complons are real. In this case the set of complonic objects determines the set of vertices of the corresponding hypergraph; with the remaining objects associated with subsets of complonic objects, adjacent to them. Thus, condition (1) in this case permits the construction of a minimal interval hypergraph, with its set of vertices formed by the set of objects i, satisfying (1), and having as its edges subsets of complonic objects, adjacent to each of the remaining objects.

In this case the construction of a minimal map leads to an ordering of the hypergraph vertices for which all edges are intervals. The ordering of an interval hypergraph is simpler than the ordering of an interval graph. We will examine the first process in §4.

However, the question arises whether all complonic objects have been given. In case they haven't it would be useful to try to artificially add new "fictitious" objects corresponding to fictitious complons. The addition of new complons is of interest not only for constructing minimal interval hypergraphs, but also as a means of analyzing situations in which the initial graph is not an interval graph. In this last case knowledge of all the complons permits their relatively easy arrangement in a non-linear structure, by the rule "place together those complons which correspond to a single mutation", whereas in terms of the initial matrix non-linear structures are unidentifiable.

3. <u>The construction of fictitious complons</u>. In this section we will give a simple constructive solution to the problem. An algorithm [23] will be given for constructing new complonal objects corresponding to fictitious complons. This algorithm terminates when all such objects have been obtained, thus permitting, in particular, the determination of whether all the complons were manifest at first by condition (1).

We will describe the algorithm in successive stages, illustrating its use on the graphs G_1 and G_2 of Fig. 9.

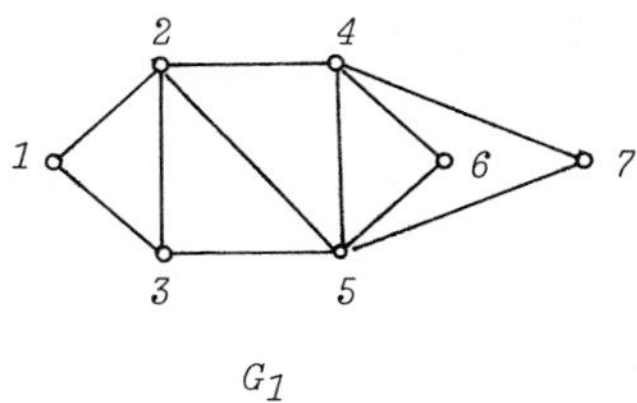

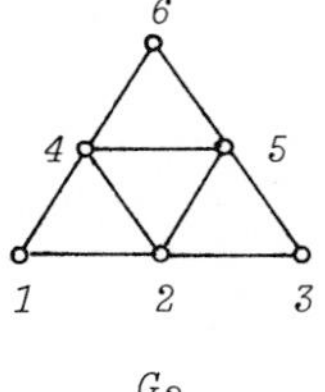

<u>Figure 9</u>

1. By means of (1) find all the complonic objects (simplicial vertices) in the initial set A.

For graph G_1 vertices $1,6$ and 7 are simplicial. For graph G_2 the complons are $1,6$ and 3.

2. Examine the subgraph I' on the set of vertices A' obtained from A by removing the vertices found in step 1.

3. Determine the set of simplicial vertices of I', and call it L.

For each $i \varepsilon L$ fix the set S_i of its complonic objects (of the initial graph) which are adjacent to i.

For G'_1 with $A' = \{2,3,4,5\}$ the set $L = \{3,4\}$, with $S_3 = \{1\}$ and $S_4 = \{6,7\}$. The graph G'_2 forms a triangle such that L coincides with A', $L = \{2,4,5\}$, with $S_2 = \{1,3\}$, $S_4 = \{1,6\}$ and $S_5 = \{6,3\}$.

4. Exclude from further consideration those $i \varepsilon L$ for which

$$S_i \cap I\langle j\rangle \neq \Phi \text{ for all } j \varepsilon I'\langle i\rangle \qquad (2)$$

Excluding the corresponding vertices from A', return to step 3. If A' is empty the algorithm terminates.

Relation (2) means there are no non-complonic objects j, adjacent to i and such that the interval I_j does not contain the complons from S_i. In this case the interval I_i, though i is complonic in I', does not contain information for the construction of new objects.

For the graph G_1 there are no objects satisfying (2). For G_2 all the objects of L satisfy (2) so the process ends, having produced no fictitious objects.

5. For each of the remaining objects $i \varepsilon L$ introduce a new object $\bar{i} \notin A$, making it adjacent with the vertices of $I\langle i\rangle \cap I\langle j\rangle$, where $j \varepsilon I'\langle i\rangle$ is selected so that $S_i \cap I\langle j\rangle = \phi$ (otherwise its selection is arbitrary). The object $\bar{i}$ corresponds to a unique fictitious complon covered in the minimal map by the interval I_i.

If, for two fictitious objects $\bar{i}$ and $\bar{j}$, $I\langle\bar{i}\rangle = I\langle\bar{j}\rangle$ then they correspond to the same complon. Keep only one of them.

For the graph G_1 we designate $\bar{3}=a$, $\bar{4}=b$. Then $I\langle a\rangle = \{a,2,3,5\}$, $I\langle b\rangle = \{b,2,4,5\}$.

6. Call this new augmented graph I, the new set of vertices A, and return to step 2. Executing step 1 again would be superfluous, as all the complonic mutants are known at this stage: They are the initial ones plus the new ones.

In actual use of the algorithm it is not necessary to build the augmented graph, since all of the same complonic objects are eliminated in A'. The introduction of fictitious objects is reflected in the changes of initial information used in steps 3 and 4.

For the graph G_1 now $S_3 \cap I\langle j\rangle \neq \Phi$ for any $j \varepsilon I'\langle 3\rangle = \{1,2,3,5\}$, since $a \varepsilon S_3$ and $a \varepsilon I'\langle j\rangle$, and similarly for S_4. Hence 3 and 4 must now be excluded from consideration.

This leaves $A' = \{2,5\}$, 2 and 5 being adjacent, so that $L = \{2,5\}$, $S_2 = \{1,a,b\}$, $S_5 = \{6,7,a,b\}$, with S_2 and S_5 intersecting $I\langle 2\rangle$ and $I\langle 5\rangle$ owing to the addition of the new objects a and b. After the exclusion of the objects 2 and 5 the process terminates.

The hypergraphs obtained for G_1 and G_2 are these:

	2	3	4	5
1	1	1	0	0
6	0	0	1	1
7	0	0	1	1
a	1	1	0	1
b	1	0	1	1

	2	4	5
1	1	1	0
6	0	1	1
3	1	0	1

Here rows correspond to complons, and columns to non-complonic mutations. The maps associated with graphs are illustrated in Figs. 10 and 11. We leave it to the reader to verify that application of the algorithm to the matrix of Fig. 2,a gives the map of Fig 2,b.

The graph G_2 is not representable by a linear map. In Section 4 we will return to the analysis of the nonlinear case, but now we prove the correctness of this algorithm.

Theorem 5. If there exists a map representing I, then the algorithm of steps 1-6 will find all its complons.

Proof. By definition S_i is a set of real complons (i.e. complons corresponding to mutations already selected in step 1), which are covered in the minimal map by the interval I_i. Clearly $i \in L$ if and only if I covers no more than one fictitious complon (which does not correspond to any $j \in S_i$), and does not cover any interval of the map with length greater than 1. Otherwise, since the removal of real complons does not

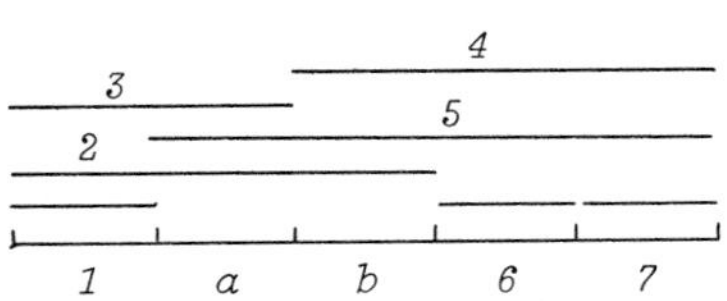

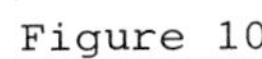
Figure 10

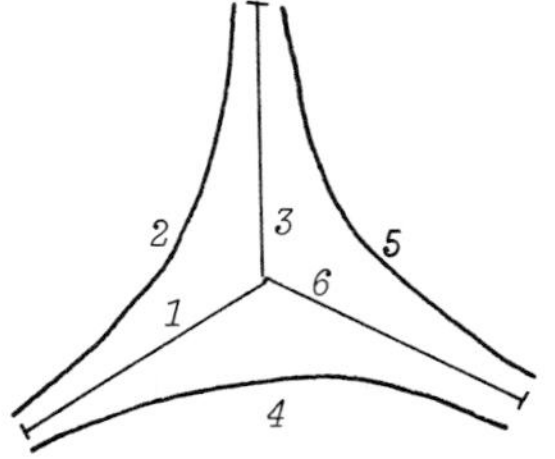

Figure 11

diminish the number of fictitious ones, repetition of the discussion in the proof of Theorem 1 would lead to the contradiction of (1), satisfied for $i \in L$ with respect to I'.

Condition (2) isolated those $i \in L$ for which I_i does not in general cover any fictitious complon, or else the covered fictitious complon is located within I_i, so that real complons are situated on both sides of it in I_i.

Indeed, if (2) is not satisfied, then there exists an interval of the map I_j ($j \in A$) such that $I_i \cap I_j \neq \Phi$, but I_j does not cover any real complon encountered in I_i. Clearly, this is possible only in the situation when the single fictitious complon is located within I_i at one end, i.e. its left (or right) end is the left (respectively, right) end of I_i.

If all the complons are real, i.e. correspond to initial objects or objects added at earlier steps, condition (2) is satisfied for all $i \in L$, since there are no fictitious objects. After removal of these i a new again-excluded set L of mutants is obtained, with intervals which cover complons and map intervals which have already been removed. This continues until all $i \in A$ are excluded, when the algorithm stops.

We now consider the case when there are fictitious complons. There then exists $i \in L$ for which (2) does not hold.

From Lemma 1, in the minimal map every natural number $k = 0, 1, \ldots$ $\ldots, m-1$ (m is the map length) is the left end of some interval I_i. In particular there is a real complon $(m-1, m)$. Let $(k, k+1)$ be the fictitious complon with maximal k, so that all complons located to its right, i.e. $(k+1, k+2)$, $\ldots$, $(m-1, m)$ are real. Consider the map interval $I_i = (k, n_k)$, where n_k is as small as possible. Clearly $n_k > k+1$, since $(k, k+1)$ is a fictitious complon.

The object i belongs to L, since I_i covers only one fictitious complon and does not cover any interval from the map of length greater than 1: All such intervals cover only real complons and are eliminated at step four of the algorithm.

Now consider an arbitrary interval I_j with right end $k+1$. Clearly $I_i \cap I_j = (k, k+1) \neq \Phi$, but $S_i \cap S_j = \Phi$, since the objects from S_i correspond to the complons $(k+1, k+2)$, $\ldots$, (n_k-1, n_k) lying to the right of I_j, and intersecting with it. Thus for i with $I_i = (k, n_k)$ i is an element of L and does not satisfy (2). Q.E.D.

The formal object $\bar{i}$ introduced for this i is such that $I_{\bar{i}}$ intersects with I_i but not with the complons $(k+1, k+2)$, $\ldots$, (n_k-1, n_k). This means the right end of $I_{\bar{i}}$ is equal to $k+1$. But the left is equal to k, since I_i does not intersect with intervals lying to the left of k. Thus $I_{\bar{i}} = (k, k+1)$ and the object $\bar{i}$ is a complon.

We have shown that the algorithm finds fictitious complons, if they exist, and stops when all complonic objects have been constructed. The theorem is proved.

From Theorem 5 it follows that the total number of complons is determined according to algorithm 1-6 only by means of the interval graph, without taking into account its possible representations, and consequently, it is fixed. From Theorem 1 the complonic objects for all minimal maps are the same. This means that no map which represents an interval graph can have fewer than m complons.

Thus the procedure described in Theorem 2.3 for identifying left and right ends of intervals of maps must lead to a minimal map. In other words the following statement holds, since both the theorem and the algorithm made use only of Lemma 1.

Corollary 2. A map is minimal if and only if each of its internal natural points is the left (and right) end for some intervals of the map.

In addition, since the length of an interval of a minimal map is equal to the number of complons covered by it, and this number is determined only by the initial complementation matrix, the following holds:

Corollary 3. A given object $i \varepsilon A$ corresponds in every minimal map to intervals of the same length.

The final solution of the uniqueness problem will be given in Section 4.2. Here, we note in conclusion that the selection of complons performed by algorithm 1-6 may be considered as characteristic of interval graphs.

Corollary 4. A graph is an interval graph if and only if the complon selection process of the algorithm, applied to the graph, produces a minimal interval hypergraph.

4. Non-linear hypergraphs and interval graphs. A map $K=\langle \mathbf{I}, I_i (i \; A)\rangle$, where $\mathbf{I} = (0,m)$ may be considered as an undirected chain joining vertices $0,1,2,\ldots,m$, with the intervals I_i $(i \varepsilon A)$ corresponding to subchains, sets of consecutively arranged edges represented by map complons.

This sets the stage for the following general definitions. Let G be an ordinary graph with the set X of edges. The set of edges $S \subseteq X$ of a chain of the graph G is a *G-interval*. The collection of G-intervals form a hypergraph on the set X which is called the *G-interval hypergraph*. The corresponding graph of the intersections of G-intervals (and graphs isomorphic to it) is the *G-interval graph*. These definitions lead to the usual "linear" graphs and hypergraphs of intervals in the particular case when G is represented by a chain.

The complementation mapping task (in the absence of linear representations) is to construct a graph G, such that the complementation graph is a graph of G-intervals. It is evident that every graph is a graph of G-intervals for a complete graph G, having sufficient vertices. Our interest is only in those graphs G which have a minimal number of edges among all G for which the given graph is a graph of G-intervals. Such a minimal graph G together with the corresponding G-intervals

($i\varepsilon A$) is naturally called a minimal map $K = <G, I_i(i\varepsilon A)>$ of the given graph, and the edges of G, complons. The edges of G are vertices of the associated G-interval hypergraph, which in this case has a minimal number of vertices.

For the notions here introduced there are a number of analogues of the statements proved in preceding sections. To formulate these analogues we will use the names and numbers of their "prototypes", adding a prime to numbers.

Lemma 1'. In a minimal map $K = <G, I_i(i\varepsilon A)>$ for every complon x there exist two intervals $I_i, I_j (i, j\varepsilon A)$ such that $I_i \cap I_j = \{x\}$.

Assuming the opposite, it is easy to see that the complon x can be removed from G by "contraction" of its intervals (by disjunctively uniting corresponding rows (and columns) of the matrix of G) without changing the intersection graph of the sets I_i (since nonempty intersections are still nonempty after the removal of x). Moreover every $I_i(i\varepsilon A)$ is a chain in the derived graph G', i.e. a G'-interval. The contradiction to the minimality of G thus derived proves the lemma.

From Lemma 1 follows

Theorem 4'. A mutant $i\varepsilon A$ is complonic if and only if it satisfies condition (1).

For non-complonic i, I_i consists of at least two neighboring complons x and y. From Lemma 1' there exist I_k, I_l, I_m, I_n such that $I_k \cap I_l = \{x\}$ and $I_m \cap I_n = \{y\}$. Hence, in particular, it follows that $I_k \cap I_l \cap I_m \cap I_n = \phi$, though $k, l, m, n\varepsilon I<i>$, which contradicts (1) and proves the theorem.

Corollary 1'. Every minimal map for a given graph has the same set of complons.

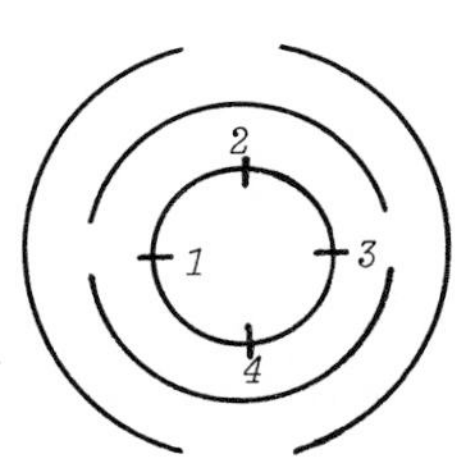

Figure 12

The selection of components in algorithm 1-6 is valid in this case also, since it is based solely on relation (1) (see the sample graph G_2 in Section 3). However, the algorithm frequently fails to select a complon. For example, on the map of Fig. 12 where the graph G is a cycle with vertices $1, 2, 3, 4$ and intervals $I_1 = \{12, 23\}$, $I_2 = \{23, 34\}$, $I_3 = \{34, 41\}$, $I_4 = \{41, 12\}$, none of the complons is real, so that algorithm 1-6 simply has nothing to catch hold on. It is necessary to make a significant modification of the algorithm so that it is applicable in this situation as well. The authors have a number of working procedures, but a strict proof of their completeness is lacking. Therefore we can formulate only the following statement, in analogy to Theorem 5.

An ordinary graph is a *tree interval graph* if it is a G-interval graph, where G is a tree.

Theorem 5'. If a graph is a tree interval graph then algorithm 1-6 will construct all complons.

It is sufficient to note that the proof of theorem 5 works for any minimal map of G for which, at each step of the algorithm, there is a sequence of real complons, associated with a fictitious complon x, which together with x constitute one of those intervals I_i, I_j for which $I_i \cap I_j = \{x\}$. A tree is clearly related to such maps because of the presence of pendant complons, trivially real by Lemma 1' (analogous to the complon $(m-1,m)$ in the proof of Theorem 5).

Corollary 4'. If algorithm 1-6 finds all complons then they determine a minimal G-interval hypergraph representing the initial G-interval graph.

Corollary 4' can be modified, considering in place of algorithm 1-6 one of its modifications that permits the analysis of situations of the type shown in Fig. 12.

Theorem 6 clearly remains true, with the replacement of the notion of an interval hypergraph by that of a G-interval hypergraph.

With respect to the characteristics of nonlinear interval graphs, results have so far been obtained only for *graphs of arcs* of circles, i.e. G-interval graphs for G defined by cycles [97,42]. Arc hypergraphs are easily characterized as follows [42].

A hypergraph G is an arc hypergraph if and only if its matrix can be brought, by permutations of its rows, to a form in which in each column the zeros or the ones are located consecutively.

We note that such a permutation corresponds to the sequential numbering of neighboring edges of a cycle of G, so that the existence of a column $k \varepsilon A$ with consecutively-placed ones means that I_k is an interval of this permutation, and a column k with consecutively-places zeros means that the complement $\bar{I}_k$ is an interval of the permutation, so that the set I_k itself is also an interval of the cycle (complementary to $\bar{I}_k$). From this remark still another criterion follows. For an arbitrary matrix Γ we construct a matrix Γ' of the same dimensions, with a column s'_k of Γ' identical with s_k of Γ if it begins with a zero. If, on the other hand, the kth column of Γ begins with a one, then s'_k is obtained from it by Boolean complementation.

Clearly Γ is an arc hypergraph if and only if Γ' is an interval hypergraph. The construction of Γ' actually means that for certain of the complons of Γ we replaced all the arcs passing through them by complementary arcs. From this it is clear that we obtain a linear map.

since a given complon does not intersect any intervals, and may be removed from the cycle G.

Graphs and hypergraphs of arcs are of definite genetic interest, since the chromosomes of a number of microorganisms (bacteria, viruses, bacteriophages, etc.) have circular form. In the first place, with their help one can establish the "closed" topological character of the corresponding recombination map. Second, by analyzing indirect "functional" data, one can arrive at the circular form of a chromosome. For example, if a circular chromosome is sufficiently well "covered" by mutations of the deletion type then, as one can easily understand, the complementation matrix corresponding to it is representable by a cyclic map. Moreover, under several reconstructions of a linear chromosome (inversions of its fragments) the deletion of several genes, which cover the inverted section, may also give the typical picture of intergenic "cyclical" complementation (see Section 5.2).

It should be said that our definition of G-interval graphs differs from precise definitions used in the literature, since for us a chain is a collection of edges and not vertices of a graph. Therefore, in particular, known properties of tree interval graphs such as triangulatedness [65a] do not hold in our case.

§4. Linear mapping algorithms

1. The Fulkerson-Gross algorithm. In this section we present algorithms for solving the linear mapping problem by ordering the objects involved.

If we have available for analysis the initial Boolean matrix $r = \|r_{ij}\|$ of the results of complementation testing ($r_{ij}=0$ if the ith and jth mutational defects do not overlap, and $r_{ij}=1$ otherwise), then we can take two routes.

In the first, we try to bring the matrix r to quasi-diagonal form and to obtain the map, as described in Theorem 1 (see Fig. 2). To bring the matrix to quasi-diagonal form we can use the idea of V. V. Shkurba, that the last complon in a map must be real, or we can use the algorithm for the orientation of the complementary graph $\bar{r}$ in order to satisfy Theorem 2.4. As a rule this approach does not work, because the initial matrix cannot be brought to quasi-diagonal form (the complementation graph is not an interval graph). Therefore we will focus on the second mapping method. First it is necessary to construct the hypergraph by the algorithm of Section 3.3 (the matrix of "complon-mutations"), and then try to order its vertices. This approach is more suitable, since the construction of fictitious complons can be carried out even if the complementation graph is not an interval graph. In this section we present the Fulkerson-Gross algorithm for the second part of the process: ordering the vertices of an interval hypergraph, i.e. the construction of a representative map (see page 37).

This algorithm may also be interpreted as an algorithm for ordering the rows of a Boolean matrix so that in each of its columns the set of ones is consecutively arranged (i.e. the matrix becomes column-linear). Terms, associated with this algorithm, finally allow us to solve the uniqueness problem for the representation of interval graphs by means of maps.

Thus, let an arbitrary hypergraph $\Gamma = (X, \{S_k\}_{k \in A})$ be given, where $X = \{x_1, \ldots, x_n\}$, $A = \{1, \ldots, N\}$. The corresponding $n \times N$ matrix of Γ has the property that its columns $S_k (k \in A)$ do not coincide with one another: $S_k \neq S_l$ for $k \neq l$, since all the edges of the hypergraph are different. We are required to find an ordering of the set X (rows of the matrix Γ) such that all edges S_k become intervals (the one appear consecutively in all columns).

The first thing that comes to mind is to successively look over the sets S_k, collecting "together" those vertices which do not violate the interval character of the ones already examined. This is precisely the idea realized in the Fulkerson-Gross algorithm, but as we shall see, its execution is associated with a number of substantial auxilliary actions.

We first select a system of subsets $S_k (k \in A)$, for which a relatively independent ordering of vertices is possible.

Consider the graph $G(\Gamma)$ with vertex set A, and edges defined by pairs kl for which $S_k \cap S_l \neq \phi$ and neither of which is a proper subset of the other (in which case we say that S_k and S_l *overlap*). The graph $G(\Gamma)$ differs from the intersection graph γ_A of hyperedges in one significant aspect: edges which connect covering and covered sets S_k are absent from it.

The components of the graph $G(\Gamma)$ are designated by $A_1, \ldots, A_p$, $(\bigcup_1^p A_s = A,\ A_s \cap A_t = \phi\ (s \neq t))$. We examine the oriented graph $D(\Gamma)$ whose vertices are the components $A_1, \ldots, A_p$, with an arc (A_s, A_t) belonging to the graph if for some $i \in A_s$ and $j \in A_t$ $S_j \subset S_i$.

Let us characterize the arcs (A_s, A_t) of the graph $D(\Gamma)$ in more detail. Consider $i \in A_s$ such that the set B of those indices $j \in A_t$ for which $S_j \subset S_i$

is nonempty. We will show that $B=A_t$. If $B\neq A_t$ then, since A_t is connected in $G(\Gamma)$, we can find an object $k\varepsilon A_t-B$ such that $S_k \cap S_j \neq \phi$ for some $j\varepsilon B$. Then clearly $S_k \cap S_i \neq \phi$ also. Since k and i belong to different components, A_t and A_s, one of the sets S_k and S_i must be a subset of the other. But $S_i \subset S_k$ is impossible since $S_j \subset S_i$, and $S_j \subset S_k$ contradicts the fact that j and k belong to the same component A_t. So $S_k \subset S_i$, i.e. $k\varepsilon B$! This contradiction proves the assertion.

We consider now the relationship of the sets $S_k(k\varepsilon A_t)$ with each of the remaining sets $S_l(l\varepsilon A_s)$. Since S_k and S_l cannot overlap each other the only remaining possibilities are that $S_k \cap S_l = \phi$ for $k\varepsilon A_t$, or that $S_k \subset S_i$ for all $k\varepsilon A_t$.

We have proved the following

Lemma 1. An arc (A_s, A_t) belongs to the graph $D(\Gamma)$ if and only if for any $l\varepsilon A_s$ one of the following holds: a) all the sets $S_k(k\varepsilon A_t)$ are contained in S_l; b) none of the sets $S_k(k\varepsilon A_t)$ intersects S_l, with the objects l for which a) holds necessarily existing.

From Lemma 1 it follows directly that the graph $D(\Gamma)$ is antisymmetric. Moreover $D(\Gamma)$ is transitive:

$$(A_s, A_t) \varepsilon D(\Gamma) \text{ and } (A_t, A_k) \varepsilon D(\Gamma) \longrightarrow (A_s, A_k) \varepsilon D(\Gamma),$$

since every $S_l(l\varepsilon A_s)$ which contains all $S_i(i\varepsilon A_t)$ also contains all $S_m(m\varepsilon A_k)$. At the same time every $S_l(l\varepsilon A)$ which does not intersect any $S_i(i\varepsilon A_t)$ also fails to intersect the sets $S_m(m\varepsilon A_k)$, since each of them is contained in at least one of the $S_i(i\varepsilon A_t)$.

Thus the graph $D(\Gamma)$ is a graph of a partial order, which characterizes the mutual disposition of the systems of subsets corresponding to different components of the overlap graph $G(\Gamma)$: Each of these systems is wholly contained in the intersection of some subset of the sets of the preceding system.

Each of the components $A_s(s=1,\ldots,p)$ of the graph $G(\Gamma)$ corresponds to a partial hypergraph $\Gamma_s = (X, \{S_k\}_{k\varepsilon A_s})$.

Theorem 1. A hypergraph Γ is an interval hypergraph if and only if each $\Gamma_s(s=1,\ldots,p)$ is an interval hypergraph.

Proof. In one direction the statement is trivial: If Γ is an interval hypergraph then for some ordering of the set X all the subsets $S_k(k\varepsilon A)$ are intervals and thus for each $s=1,\ldots,p$ the subsets $S_k(k\varepsilon A_s)$ are intervals since $A_s \subseteq A$.

Now let all the Γ_s be interval hypergraphs. We are required to show that Γ is also an interval hypergraph, i.e. that the intersection of nonempty sets of the admissible orderings of the hypergraphs Γ_s for each $s=1,2,\ldots,p$, is nonempty (it consists of admissible orderings of the hypergraph Γ). This follows from the fact that the sets S_k for $k\varepsilon A_s$ affect only a local segment of the set X, the ordering of which does not depend on the ordering of the components of the graph $D(\Gamma)$ which precede or are unconnected with A_s.

We will give a strict proof.

For clearness the argument will be carried out in terms of the matrices Γ_s corresponding to the hypergraphs $\Gamma_s (s=1,\dots,p)$. Then the matrix of the hypergraph Γ can be represented as a sequence of submatrices $\Gamma_s(s=1,\dots,p)$: $\Gamma = (\Gamma_1, \Gamma_2, \dots, \Gamma_p)$ ordered one after another in accord with the graph $D(\Gamma)$. Thus Γ_p must correspond to a minimal component of $D(\Gamma)$ (i.e. a vertex from which there are no outgoing arcs).

We first note that in an admissible ordering identical rows may be placed together. The exclusion of rows (i.e. vertices of X) cannot make an interval graph into one that is not. Therefore we can exclude one of the identical rows, find an admissible ordering of the resulting matrix, and then insert the excluded row next to its twin, since this does not violate the linearity of the matrix. Repeating this operation we obtain an admissible ordering of the rows in which identical rows are situated together.

Now consider the submatrices $\Gamma^s = (\Gamma_1, \dots, \Gamma_s)$ formed by the first s components of the graph $D(\Gamma)$ $(s=1,\dots,p)$. Clearly the matrix Γ_s corresponds to the minimal A_s in the sense of the partial order $D(\Gamma^s)$.

The proof will proceed by induction on s. For $s=1$ $\Gamma^1 = (\Gamma_1)$, but Γ_1 is column-linearizable by agreement. Suppose Γ^{s-1} is column-linearizable in such a way that identical rows appear consecutively. We will show that Γ^s is also column-linearizable.

We arrange the rows of Γ^s so that Γ^{s-1} is column-linear. It will have the form $\Delta^s = (\Delta_1,\dots,\Delta_{s-1},\Delta_s)$, where $\Delta^{s-1} = (\Delta_1,\dots,\Delta_{s-1})$ is column-linear. In the submatrix Δ_s, let i_1 be the index of the first row (from the top) in which a one appears, and let i_2 be the last row in which a one appears, so that all rows above i_1 and below i_2 contain only zeros. We will show that in the matrix Δ^{s-1} all the intermediate rows i, with $i_1 \leqslant i \leqslant i_2$, are identical.

In $D(\Gamma)$ A_s is minimal among $A_1,\dots,A_s$, so that every $A_t (1 \leqslant t < s)$ either precedes A_s or is not connected with it. If A_t is not connected with A_s then in Δ_t the rows i_1 and i_2 are zero (since they are zero in Δ_s) and are consequently identical. If, on the other hand, A_t precedes A_s, by Lemma 2 the rows i_1 and i_2 in Δ_t are also identical. This means that in Δ^{s-1} the rows i_1 and i_2 coincide, and consequently, by the induction hypothesis, so do all intermediate rows.

Thus every permutation of the rows of Δ_s lying between i_1 and i_2 maintains the linearity of the matrix Δ^{s-1}, but only such permutations (by the definition of i_1 and i_2) are necessary to bring Δ_s to column-linear form. Selecting one of these permutations we obtain a column-linear form for the matrix Γ^s, which proves the theorem.

The foregoing proof shows that to order Γ it is sufficient to order each Γ_s individually, in the sequence which corresponds to the graph of components $D(\Gamma)$.

It remains to give the algorithm for ordering the rows of a matrix Γ_s corresponding to a component A_s.

We first examine the case in which the component $G(\Gamma_s)$ has only three vertices, i.e. the matrix has three columns s_1, s_2, s_3.

Consider now the scalar products of these columns. Clearly $s_i \cdot s_i$ is the number of ones in column s_i, $s_i \cdot s_j = s_i \leftrightarrow S_i \subseteq S_j$, $s_i \cdot s_j = 0 \leftrightarrow S_i \cap S_j = \phi$ $(i,j=1,2,3)$.

Since the columns s_i are contained in one component, at least two pairs of corresponding edges overlap. Let these be, say, $S_1 \cap S_2 \neq \phi$ and $S_2 \cap S_3 \neq \phi$. In this case we will call the ordered triplet s_1, s_2, s_3 an *overlapping triplet*. Of course $s_1 \cdot s_2 \neq 0$ and $s_2 \cdot s_3 \neq 0$.

Two cases are possible

$$s_1 \cdot s_3 \quad \min(s_1 \cdot s_2, s_2 \cdot s_3) \tag{3}$$

or

$$s_1 \cdot s_3 \quad \min(s_1 \cdot s_2, s_2 \cdot s_3) \tag{4}$$

It is easy to show that in the linear form of a hypergraph matrix each of these cases corresponds to a specific mutual arrangement of the subcolumns of ones (i.e. intervals S_1, S_2, S_3) following one another. Case (3) corresponds to the ladder-like arrangement of Fig. 13, and case (4) to the "inverted hump" of Fig. 14. Here the sections of consecutive ones of corresponding columns are shown as vertical segments, with zeros not shown.

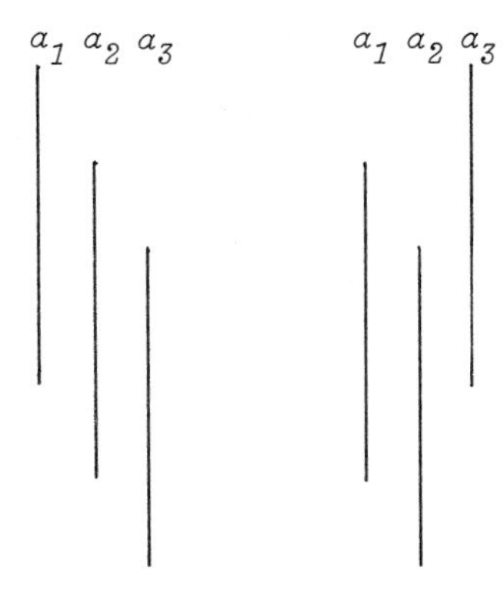

Figure 13 Figure 14

The situation of Fig. 13 is obtained by placing $s_1 \cdot s_1$ ones together in the left column, and then $s_2 \cdot s_2$ ones in the second column with an intersection of $s_1 \cdot s_2$ ones with the first column, and then $s_3 \cdot s_3$ ones in the third with an intersection of $s_2 \cdot s_3$ ones with the second column. The analogous situation obtains for Fig. 14 with respect to (4).

We emphasize that there are two possibilities for the placement of the ones in the second column: lower than those in the first, as in Figs. 13 and 14, or higher. Once this selection is made the position of the ones in the third column is uniquely determined from (3) and (4), since the information about the scalar products determines the relative positions of the segments up to reversals.

Having fixed the relative positions of the ones, it remains to determine the number of common ones in the first and third columns. If this quantity is not equal to $s_1 \cdot s_3$ then as a result of the uniqueness of the construction we can say that Γ_s is not linearizable. However, if equality holds, we can conclude that any ordering connected with S_1, S_2, S_3 as described, is admissible.

In the case when contains more than three columns, the construction is easily extended to all possible overlapping triplets.

For this it is sufficient to construct the spanning tree of the component A_s of the overlap graph. As is shown in the appendix (page 187) the spanning tree of a component (which, itself, may not be specified in advance) is determined by the following procedure. Beginning with an arbitrary vertex, examine all those adjacent to it, and for each of these new vertices in succession, those adjacent to it, and so forth, until no new vertices can be found (which marks the exhaustion of that component). It is important that edges of the spanning tree be formed only between a given vertex (at any stage) and new adjacent vertices.

Once the spanning tree is formed, the overlapping triplets are those triplets whose members are successively connected by edges of this tree. These are enough, since in any other overlapping triplet the first two columns are contained in some overlapping spanning-tree triplet, so that their relative disposition is uniquely determined by the spanning tree.

The process begins with the ordering of an arbitrary overlapping spanning-tree triplet of columns. In the general stage (of the algorithm) we consider a new overlapping triplet, whose first two columns are already ordered, so that positioning of the ones in its third column is uniquely determined by the associated scalar products. After the ones of the third column are positioned in accord with Fig. 13 or Fig. 14, it is necessary to verify that there is no contradiction in the configuration obtained so far. To do this the third column (of this latest triplet) is compared with each of the previously ordered columns, with respect to their scalar products and their common ones, more precisely, the number of vertices in the intersection of the corresponding hyperedges.

As a result of the uniqueness (for a given skeleton) of the configuration obtained so far, inequality in any of the comparisons means the matrix Γ_s is not column-linearizable. If equality is found in all the comparisons the admissibility of the ordering so far obtained is supported, and the next overlapping triplet (with one new column) is selected for examination. The process ends when all columns have been examined.

We obtain as a result a mutual disposition of the intervals $S_k (k \varepsilon A)$ which easily allows the ordering of the vertices, or in general, the transition to a minimal map by observation of the interval order as was done in Section 2.1.

Thus the Fulkerson-Gross algorithm for ordering the vertices of a hypergraph consists of the following steps:

1. Build the overlap graph $G(\Gamma)$.

2. Through the construction of spanning trees, separate out the components A_s of the graph $G(\Gamma)$ $(s=1,2,\ldots,p)$. The spanning tree of each component A_s is, at once, the spanning tree of the graph $G(\Gamma_s)$ for the partial hypergraph Γ_s associated with A_s.

3. Build the graph $D(\Gamma)$ of the partial order of the components A_s and renumber the A_s so that $(A_s, A_t) \varepsilon D(\Gamma) \longrightarrow s<t$.

4. Set s to 1.

5. Calculate the matrix of scalar products of all pairs of columns of the Γ_s matrix.

6. Using the spanning tree of the component A_s order the ones in the columns of Γ_s, by comparing their scalar products (page 55). If these comparisons lead to a contradiction then the desired ordering does not exist.

7. Increment s by 1. If $s \not> p$ go to step 5. Otherwise, stop. The ordering of ones in the columns of the hypergraph matrix Γ is consistent. The corresponding ordering of the rows of Γ is the one disired.

For example, let the matrix Γ_s have the following form:

	1	*2*	*3*	*4*	*5*	*6*
1	1	1	0	0	1	0
2	1	1	1	1	0	0
3	1	0	0	0	0	0
4	0	0	1	0	0	1
5	0	1	1	1	0	1
6	1	1	0	1	1	0

The component of the overlap graph of columns is shown in Fig. 15,a. Beginning with vertex *5* we obtain the spanning tree of the graph shown in Fig 15,b.

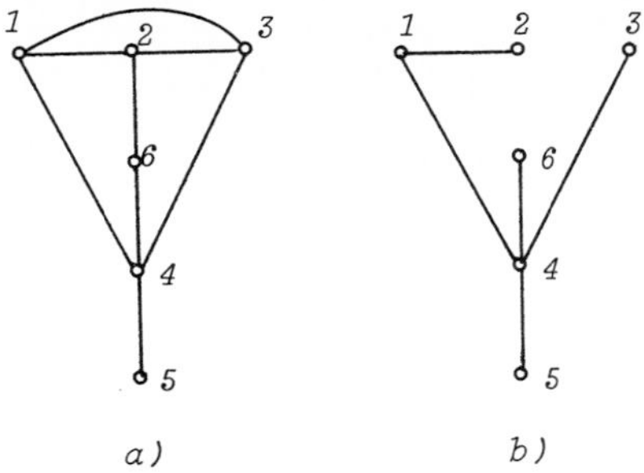

Figure 15

The matrix of scalar products is:

	1	2	3	4	5	6
1	4	3	1	2	2	0
2		4	2	3	2	1
3			3	2	0	2
4				3	1	1
5					2	0
6						2

We will successively examine the overlapping triplets of the spanning tree, beginning with *214*. Condition (4) is satisfied for this triplet, so that we obtain the "inverted hump" configuration.

2	1	4
1		1
1	1	1
1	1	1
1	1	
	1	

The number of common ones in columns *2* and *4* is three, and $s_2 \cdot s_4 = 3$ so that we can continue the process.

Column *3* is added (from the triplet *143*, satisfying condition(3)):

2	1	4	3
			1
1		1	1
1	1	1	1
1	1	1	
1	1		
	1		

Comparing $s_2 \cdot s_3 = 2$ and $s_1 \cdot s_3 = 1$ with the numbers of ones in the associated pairs of columns shows no contradictions in the configuration obtained so far.

Adding column *6* from the triplet *146* satisfying condition (3):

2	*1*	*4*	*3*	*6*
			1	1
1		1	1	1
1	1	1	1	
1	1	1		
1	1			
	1			

Comparison of the values $s_2 \cdot s_6 = 1$, $s_1 \cdot s_6 = 0$, $s_3 \cdot s_6 = 2$ with the numbers of common ones again supports the admissibility of the ordering.

Finally, column *5* is added, for example, from the triplet *645*, satisfying condition (3). This gives:

2	*1*	*4*	*3*	*6*	*5*
			1	1	
1		1	1	1	
1	1	1	1		
1	1	1			1
1	1				1
	1				

Having convinced ourselves of the admissibility of this configuration we construct the desired row ordering. From the examination of columns *1* and *2* the first position must be occupied by the row whose number is common to the two sets $\{4,5\}$ and $\{3,4\}$, i.e. *4*; then *5*; and *3* last. From column *4* it is clear that row *1* is next-to-last. Information about columns *3* and *6* shows how to place the remaining rows. The final ordering (permutation) of the rows is *4,5,2,6,1,3* (of course the reverse: *3,1,6,2,5,4* is also admissible).

2. The uniqueness theorem. We first note that the process just described provides a proof for Theorem 3.2, since it uses only information about the scalar products of columns (including scalar squares, giving the number of ones in a column), on which formulas (3) and (4) are also based; and these uniquely determine (except for reversals) the ordering of the columns k for a component A_s $(s=1,\ldots,p)$.

From this it follows that in the linear case identical matrices $\|s_i \cdot s_j\|$ of scalar products must lead to identical configurations on ones, so that the initial matrices can differ in the number of zero rows and in the ordering of the other rows. If, however, a matrix $\|s_i \cdot s_j\|$ does not allow an admissible configuration this signifies that, whatever the initial hypergraph that led to it, it is not linearizable.

The uniqueness properties of the Fulkerson-Gross method permit us to conclude the solution of the uniqueness problem for complementation mapping.

Theorem 2. If Γ is a minimal linear hypergraph for which the overlap graph contains p components, then there are exactly 2^p distinct admissible orderings (depending on selection of one of two mutually inverse orderings of the vertices corresponding to each component).

Proof. Suppose, as in Theorem 1, the graph $G(\Gamma)$ has p components $A_1, \ldots, A_p$ with a corresponding partition of the matrix $\Gamma = (\Gamma_1,\ldots,\Gamma_p)$, where the enumeration of the components corresponds to the partial ordering given by the graph $D(\Gamma)$.

Let $= (\Delta_1, \ldots, \Delta_p)$ be a (column)-linear form of Γ obtained by permuting the rows i. As in the proof of Theorem 6, we argue that the non-zero rows of the matrix Δ_p form an interval $[i_1,i_2]$ of the ordering which "covers" a set of identical rows of the matrix $\Delta^{p-1}=(\Delta_1,\ldots,\Delta_{p-1})$. This means that in Δ_p all the non-zero rows are distinct, since from the minimality of Γ no two rows coincide. As a consequence of the Fulkerson-Gross algorithm, this means the admissible ordering of the interval $[i_1,i_2]$ of Δ_p is uniquely defined up to reversals. Actually, in the case when all rows are different an arbitrary admissible configuration of the ones in the intervals uniquely determines the range of the vertex associated with each of the rows obtained. Therefore a unique ordering of the rows corresponds to each of the two possible mutual dispositions of the intervals.

We now examine the matrix $\bar\Delta = (\bar\Delta_1, \ldots, \bar\Delta_{p-1})$obtained by eliminating from Δ all rows of the half-open interval $(i_1,i_2]$ (excepting i_1), as well as the trivial submatrix $\bar\Delta_p$ with the single remaining non-zero row i_1. This matrix is clearly linear. Moreover, in $\bar\Delta$ all rows are dis tinct. Identical rows of $\bar\Delta$ not equal to i_1, cannot occur, since such rows would also be identical in Δ, because of their zero extensions in Δ_p. But rows which coincide with i_1 in $\bar\Delta$ also cannot occur, since this contradicts the minimality of the initial hypergraph. In fact, every interval $S_k(k\notin A_p)$, because of Lemma 1 and the minimality of A_p in $D(\Gamma)$, either contains all $S_l(l\varepsilon A_p)$ (among them also the row i_1), or does not intersect at all with them and does not contain i_1. The intervals S_k $(k\notin A_p)$ must behave in just this way also with respect to a row i which coincides in $\bar\Delta$ with i_1. Moreover, in Δ_p this row i is zero, since it cannot coincide with i_1 in Δ. This means the row i can be removed from Δ without affecting the intersection relation of the sets $S_k(k\varepsilon A)$, but this contradicts Corollary 2 on the minimality of Γ.

The components of the graph $G(\bar\Delta)$ coincide with components of the graph $G(\Delta)$ $A_1, \ldots,A_{p-1}$, and the graph $D(\bar\Delta)$ coincides with $D(\Delta)$ with the exclusion of A_p. In fact, the intersection relation of columns is unchanged if (as was done) we exclude duplicate rows.

Thus, for the matrix $\bar{\Delta}$ we can repeat the argument that only two mutually inverse permutations of the interval $[i_1, i_2]$ are possible, and that these are controlled by the minimal matrix, in the sense of $D(\Gamma^{p-1})$ (in this case Γ_{p-1}). Repeating this procedure p times proves the theorem.

The above proof provides not only the number of possible orderings of a minimal graph, but also a means of obtaining them from any specific ordering.

To do this it is necessary to select from s given minimal map those intervals $S=(k,n_k)$ (not necessarily map intervals) having the property that each map interval which intersects with S either is contained in it or contains it. Such an interval S corresponds to a component of the intersection graph, as is required by the proof of Theorem 2. Therefore, reversing the order of the complons of S leads to a new minimal map.

All the possible minimal maps can be obtained from a given minimal map by successive applications of this technique.

3. Admissible orderings of linear matrices. In this section we examine linear matrices. We referred to these matrices in connection with the recombination mapping problem (page 18). However, such matrices can appear naturally also in other situations -- anytime connections between objects reflect their linear order. Connections between objects that are near one another in the ordering must exceed connections between objects more distant from one another. In particular, linear matrices of connections arise repeatedly in the study of complementation maps. Thus, the complementation matrix for non-overlapping mutational distortions (the graph of non-covering intervals) is linear, as is the matrix of connections between separate sections of a map, i.e. complons (Theorem 3.1).

As it turns out, the theory of linear matrices [13] falls naturally within the theory of graphs and hypergraphs of intervals.

Thus, suppose that $\|a_{ij}\|$ is a matrix of connections between objects. We will say that it is *linear* (with respect to rows) if and only if

$$a_{ij} \geqslant a_{ik} \text{ for } i \leqslant j \leqslant k \text{ or } i \geqslant j \geqslant k. \tag{5}$$

In some situations the values a_{ij} characterize the distance between i and j: The nearer i and j are, the smaller a_{ij} is, since the interpretation of a_{ij} as a connection leads to a reverse dependence (smaller for a reduced connection between i and j). In such situations the notion of linearity (5) is naturally modified by substitution of the inequality $a_{ij} \leqslant a_{ik}$ for $a_{ij} \geqslant a_{ik}$. We will carry out all subsequent constructions using the relation (5), but their modifications in the case of distance is trivial. It is necessary only to use the relation "less-than" in place of the relation "greater-than."

An ordering $p = (i_1, \ldots, i_N)$ of objects, which brings a matrix to linear form will be said to be *admissible*. We will formulate admissibility critera for orderings.

Suppose the elements of the ith row of a matrix $\|a_{ij}\|$ take on m_i distinct values $a_1, \ldots, a_{m_i}$, with $a_1 > a_2 > \ldots > a_{m_i}$. We define

$$R_1^i = \{j \mid a_{ij} = a_1\}, \quad R_2^i = \{j \mid a_{ij} = a_2\}, \quad \ldots, \quad R_{m_i}^i = \{j \mid a_{ij} = a_{m_i}\}.$$

The set $R_k^i (i \varepsilon A,\ k=1, \ldots, m_i)$ thus consists of those objects having identical connections with i (monotonically decreasing with increase of k). Consequently, the ith row of the matrix $\|a_{ij}\|$ induces an ordered partition (ranking) $R^i = (R_1^i, \ldots, R_{m_i}^i)$ of the set A.

Theorem 3. An ordering $p = (i_1, \ldots, i_N)$ is admissible if and only if all the sets

$$S_k^i = \bigcup_{j=1}^{k} R_j^i \quad (i \varepsilon A,\ k=1, \ldots, m_i) \tag{6}$$

are its intervals.

Proof. From (6) $S_k^i = \{j \mid a_{ij} \geqslant a_k\}$, so that S_k^i consists of objects with which i has the largest connection. But in the linear form of the matrix columns corresponding to the largest elements of the row are situated consecutively. Therefore from the admissibility of the ordering it follows that the sets S_k^i are its intervals.

If, on the other hand, for some ordering p all the S_k^i are intervals then the simultaneous permutation of rows and columns of the matrix $\|a_{ij}\|$ in accord with p brings it to linear form. Thus in the ith row the maximal elements a_{ij} with $j \varepsilon R_1^i$ are situated next to a_{ii} (since by agreement $S_1^i = R_1^i$ is the interval around them), followed (in order of size) by the elements a_{ij} with $j \varepsilon R_2^i$ (since $S_2^i = R_1^i \cup R_2^i$ is the interval ...), etc. The theorem is proved.

Corollary 1. The matrix $\|a_{ij}\|$ is linearizable if and only if the system $(A, \{S_k^i\}_{i \varepsilon A,\ k=1 \ldots, m_i})$ is an interval hypergraph.

Thus the problem of constructing an admissible ordering is transformed into the problem of ordering an interval hypergraph, and it may be solved by the method of Fulkerson and Gross.

It is useful to consider another construction, connected with a specific characteristic of swuare linear matrices. For example, in the first row of a linear matrix the elements are decreasing, while in the last they are in creasing. In order to formulate such properties we veral notions [13].

We say that an ordering p *includes* an ordered partition R if and only if $iRj \longrightarrow ipj$, i.e. p is obtained from R by the additional ordering of the objects internal to the classes of R. A set S is called an *interval of the ordered partition* R, if and only if it is an interval of some ordering contained in R.

The set S is said to *cover* an object k in the ordered partition R, if there exist $i,j \varepsilon S$, such that iRk and kRj. Then an interval of an ordered partition is a set which contains all objects covered by it.

In other words, a set S is an interval of an ordered partition $R = (R_1, \dots, R_m)$, if and only if for some $s,t (1 \leqslant s \leqslant t \leqslant m)$

$$S = P_s \cup R_{s+1} \cup \dots \cup R_t \cup P_{t+1},$$

where $P_s \subset R_s$, $P_{t+1} \subset R_{t+1}$. We will say that the ordered partition

$$R' = (R_1, \dots, R_{s-1}, R_s - P_s, P_s, R_{s+1}, \dots, R_t, P_{t+1}, R_{t+1} - P_{t+1}, R_{t+2}, \dots, R_m)$$

is obtained by *superimposing the interval* S on R.

Property 1. If the ordering $p = (i_1, \dots, i_N)$ is admissible and $i_1 \varepsilon R_1^l$ for some $l \varepsilon A$, then p includes the ordered partition R^l. In particular, p includes the ordered partition R^{i_1} induced by the i_1th row of the matrix $\|a_{ij}\|$.

Property 2. An admissible ordering includes an ordered partition R if and only if it includes the ordered partition R', obtained by superimposing on R its interval S which coincides with one of the sets (6).

Property 3. Suppose $R = (R_1, \dots, R_m)$ is obtained from the ordered partition $R^l = (R_1^l, \dots, R_m^l)$ (for some $l \varepsilon A$) by successive superimposings of some of its intervals (6) (it is possible that $R = R^l$). Suppose further that for some $i \varepsilon A$, $k = 1, \dots, m_i$ the set S_k^i is not an interval of R, so that there exists an object $s \notin S_k^i$ which is covered by S_k^i. Then there do not exist admissible orderings beginning with any object which precedes s in R.

Proof 13 . To prove property 1 it is sufficient to note that if $l = i_k$, then $a_{li_1} \leqslant a_{li_2} \dots \leqslant a_{li_{k-1}}, a_{li_{k+1}} \geqslant \dots \geqslant a_{li_N}$, but since $i_1 \varepsilon R_1^l$ then also $i_{k-1} \varepsilon R_1^l$, so that in fact

$$a_{li_1} = \dots = a_{li_{k-1}} \geqslant a_{li_{k+1}}$$

Thus the elements $a_{li_1}, \dots, a_{li_N}$ form a non-increasing sequence, and this means that R^l is contained in $p = (i_1, \dots, i_N)$.

We now prove property 2. If the admissible ordering includes R' then clearly it includes R also. If, on the other hand, the admissible ordering p includes R but does not include R' then this means that in p some element of the set P_s precedes some element of the set $R_s - P_s$, or some element of the set $R_{t+1}-P_{t+1}$ precedes at least one element of P_{t+1}. But then the set S is not an interval of the ordering p and, by Theorem 1 p is not admissible.

We now prove property 3. First, note that the sets $S_k = \bigcup_{j=1}^{k} R_j$ $(k=1,\ldots,m)$ are intervals of the admissible ordering. Suppose that R is obtained from R^l by superimposing some intervals of the form (6). Consider R', obtained from R by superimposing still another interval S of the form (6). Clearly only the sets

$$S'_s = \bigcup_{j=1}^{s-1} R_j \cup (R_s-P_s),\ S'_{t+1} = \bigcup_{j=1}^{t} R_j \cup P_{t+1}$$

are not unions of the first classes of the ranking R, so that it is necessary to verify whether they are intervals of the admissible ordering. But

$$S'_s = \bigcup_{j=1}^{s} R_j - S = S_s - S, \qquad S'_{t+1} = \bigcup_{j=1}^{t+1} R_j - S = S_{t+1} - S,$$

with S_s and S_{t+1} and S all being intervals of the admissible ordering. But then S'_s and S'_{t+1} are also intervals, as the differences of intervals that are not included in each other. Thus we have shown that the sets $S_k = \bigcup_{j=1}^{k} R_j$ $(k=1,\ldots,m)$ are intervals of the admissible ordering (if it exists).

Now suppose that in R the object t precedes the object s, for which $s \notin S_k^i$ for some $i \in A$, $k=1,\ldots,m_i$, with $s_1, s_2 \in S_k^i$ such that $s_1 R s$ and $s R s_2$. We suppose that an admissible ordering p, beginning with t exists.

There are two possible cases for the arrangement of the objects t, s and l in p: a) $p = (t\ldots l\ldots s\ldots)$ or b) $p = (t\ldots s\ldots l\ldots)$. Consider case a). All of the sets $S_k = \bigcup_{j=1}^{k} R_j$ are intervals of p, so that $l,t \in S_{k_1}$; $l,t,s_1 \in S_{k_2}$; $l,t,s_1,s \in S_{k_3}$; $l,t,s_1,s,s_2 \in S_{k_4}$, with $k_1<k_2<k_3<k_4$ by assumption. From this it follows that the objects s_1 and s_2 are situated in p as follows: $p = (t\ldots l\ldots s_1\ldots s\ldots s_2\ldots)$. But then S_k^i is not an interval of p, so that p is not admissible, by Theorem 1.

Case b) goes similarly and leads to the same contradiction, so that property 3 is proved.

From properties 1-3 comes the following algorithm for discovering all admissible orderings. For each $l \in A$ the algorithm finds all admissible orderings which begin with l, or else shows that there are none.

Examined candidates for first place are stored in the set L. At first L is empty.

For each $l \notin L$ examine the ordered partition R^l and some arbitrary S_k^i of the form (6). If S_k^i is not an interval of R^l then by property 3 for any $s \notin S_k^i$ covered by R^l there does not exist an admissible ordering for any objects t which precede s. Place these t in L and return to examine a new R^l for $l \notin L$. If S_k^i is an interval of R^l, then proceed to examine $R^{l\prime}$, obtained by superimposing S_k^i on R^l, and a new set S of the form (6). By properties 1 and 2 all admissible orderings beginning with elements of R_1^l, if they exist, are contained in $R^{l\prime}$. Successively repeating this superpositioning procedure for all sets $S_k^i (i \in A,\ k=1,\dots,m_i)$ we either arrive at an ordered partition R, for which one of the sets of the form (6) is not an interval (in which case we place new objects t in L as above and proceed with a new $l \notin L$), or at an ordered partition R for which all the sets S_k^i of the form (6) are intervals, so that all the orderings contained in R are admissible[†]. In this last case place all elements of the set R_1 in L and return to the examination of R^l for $l \notin L$.

The process terminates when $L = A$. L may be filled more quickly if we keep in mind that the reverse ordering of any admissible ordering p is also admissible. Having obtained at least one admissible ordering, we can also make use of the uniqueness theorem (Theorem 2) to discover the remaining orderings.

Admissible orderings frequently do not exist. We consider the following generalization of this notion. Every ordered partition $R = (R_1, \dots, R_m)$ of the set A induces a partition of the matrix $\|a_{ij}\|$ into submatrices (cells) $B_{st} = \|a_{ij}\|$ $(i \in R_s,\ j \in R_t)$. We say that a cell B_{st} is *not greater than* a cell B_{sr}, if each element of each row of B_{st} does not exceed the minimal element of the corresponding row of the cell B_{sr}:

$$B_{st} \leqslant B_{sr} \leftrightarrow \forall i \in R_s \left[\max_{j \in R_t} a_{ij} \leqslant \min_{j \in R_r} a_{ij}\right].$$

The fact that $B_{st} \leqslant B_{sr}$, means that the closeness of the connections of objects of class R_s with objects of class R_t never exceeds the closeness of their connections with objects of R_r.

An ordered partition $R = (R_1, \dots, R_m)$ of a set A is said to be *admissible*, if the matrix of cells $\|B_{st}\|$ has linear form, in the sense of the relation just introduced:

$$B_{st} \leqslant B_{sr}, \text{ iff } s \leqslant r \leqslant t \text{ or } s \geqslant r \geqslant t. \tag{7}$$

A set $S \subseteq A$ is said to be *closed* (with respect to the system of subsets S_k^i of form (6), if it contains, along with all $i \in S$, all sets S_k^i $(k=1,\dots,m_i)$ which do not contain S.

Theorem 4. An ordered partition is admissible if and only if all its classes are closed and all the sets S_k^i of the form (6) are its intervals.

[†]Of course, as in Section 1, only one component of the overlap graph of sets S_k^i is being considered. The other components correspond to subsets of objects, and are considered separately.

The proof of this theorem and of the algorithm (based on it) for finding admissible ordered partitions with a maximal number of classes are given in [13].

This result is connected with the problem of approximating real data by means of linear matrices. If an admissible ordering does not exist then an ordered partition with a maximal number of classes may be considered as an approximating structure, which characterizes the desired ordering up to permutations of objects within classes.

Probably a more natural theory for approximating data by graphs and hypergraphs of intervals can be developed within the framework of the graph approximation idea presented in Chapter 2. However we will not present any theoretical or experimental results in this direction. The particular problem of approximating a partial order by an interval graph contained in it, was examined recently by F. Aleskerov [1a].

§5. Examples of structural analysis of genetic systems

1. The use of deletions and polar mutations. The majority of mutations, both spontaneous and induced, occur as simple replacements of nucleotides in DNA, leading to corresponding replacements of amino acids in proteins.

The use of only such point mutations for constructing finely-detailed maps is not suitable in recombination mapping because of the necessity of producing a very large number of crosses, nor in complementation mapping because they do not provide information about the meshing of text fragments.

Very frequently mapping is made easier by the use of deletions, the loss of genome fragments of various lengths. Operationally (in genetic analysis), deletions are characterized by the following two properties: a) unlike point replacements, which permit the reverse mutations (reversions) to return the organism to the norm, deletions are stable and do not revert; b) deletions, depending on their lengths, are not able to recombine with some point mutations [30,43,48]. At the molecular level property a) is explained by the infinitesimally small probability of making up for the loss of an entire series of nucleotides, at times of great length, by inserting the lost text. Property b), at the molecular level, means that a deletion which overlaps a point mutation cannot recombine with it.

These properties make for wide (and highly successful) use of deletions both in recombination and complementation gene mapping.

The classic example of the use of deletions in constructing a recombination map is the work of S. Benzer [3,55] (see also page 18) in analyzing the fine structure of the rII locus in the bacteriophage T4, a virus of the intestinal bacterium *E. coli*. The bacteriophage destroys an *E. coli* cell within an hour, but mutations of this locus can significantly shorten this period (hence, this locus and its mutations are called *rapid-lysis*).

The collection of S. Benzer included more than 2400 mutations of this locus. The usual recombination approach would require an astronomical number of paired crossings. Therefore Benzer chose mutations able to revert to the norm (a priori, point mutations). He did recombination mapping by the usual paired crossings for several dozens of these mutations. Then he was able to select 32 deletions (according to their inability to revert) and to cross them with the point mutations which were already located on the map. Each deletion was associated with its list of point mutations with which it gave no recombinant progeny. In other words, the point mutations gave the vertices, and the deletions, the edges of the interval hypergraph. It turned out that this graph was divided into a system of imbedded hypergraphs Γ_s, i.e. the graph $G(\Gamma)$ had several components A_s (see Section 4.1) (Fig. 16a). Benzer first examined seven of the longest deletions, which showed a step-like arrangement. Then, crossing the remaining point mutations with these deletions, it was not difficult to establish on which "step" (i.e. on which of the distinguishing segments for neighboring deletions) their defects were located. A mutation is located in the segment distinguishing deletion i from deletion $i+1$ if it is not recombinant with the former, but is with the latter.

In this way all the mutations were divided among the seven segments formed by the deletions.

In succeeding stages Benzer carried out a similar procedure, working with deletions at succeeding "levels", localized within individual segments at higher levels (see Fig. 16a).

As a result he was able in three steps to divide the rII region into 47 deletion segments, "equivalent" to complons of the deletion intersection graph (see Section 3.2), and also was able to distribute the whole collection of mutations of the rII locus among those segments (see Fig. 16b). The arrangement of the point defects within these segments was determined by the usual recombination analysis.

The work of Benzer is a model of the use of qualitative analysis in mapping. It is no accident that it has attracted the interest of mathematicians to graphs and hypergraphs of intervals. Owing to the successful selection of deletions Benzer did not have to solve the algorithmic problems of ordering graphs and hypergraphs of intervals, but the stimu-

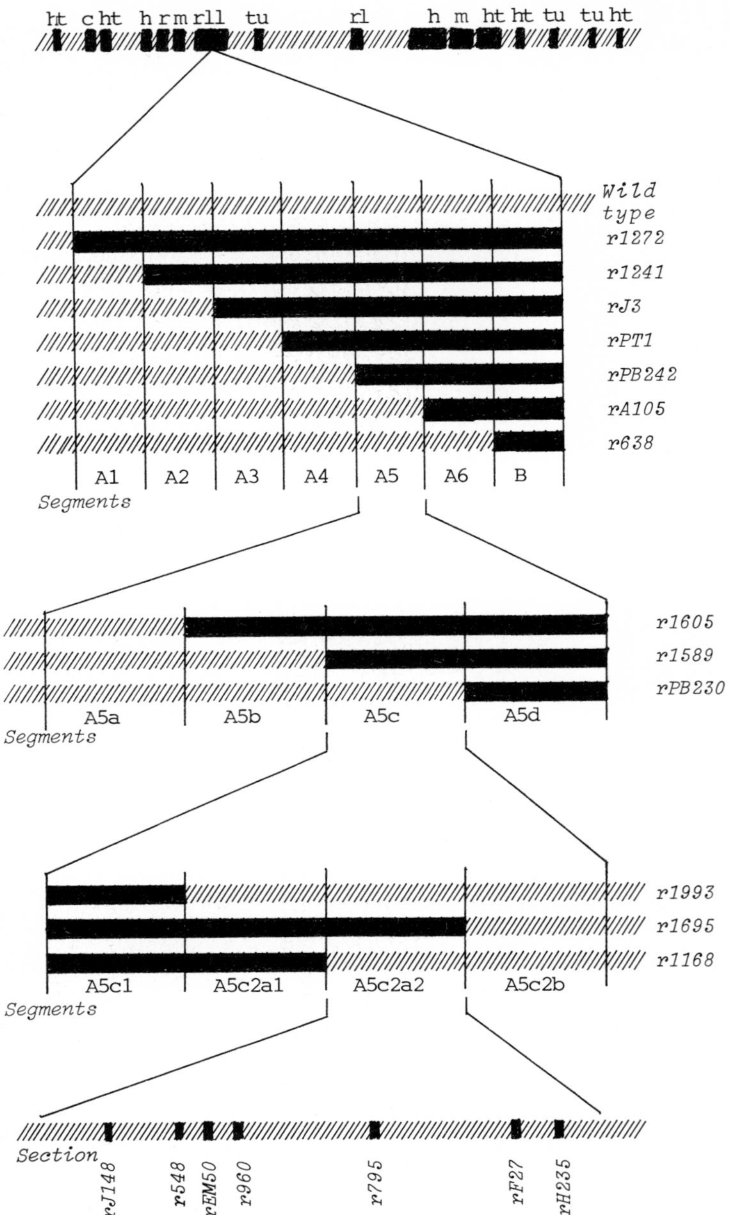

Figure 16a. The mapping of mutations in the rII locus of the bacteriophage T4 [3,55]. The three-stage division of the locus into segments by means of overlapping deletions.

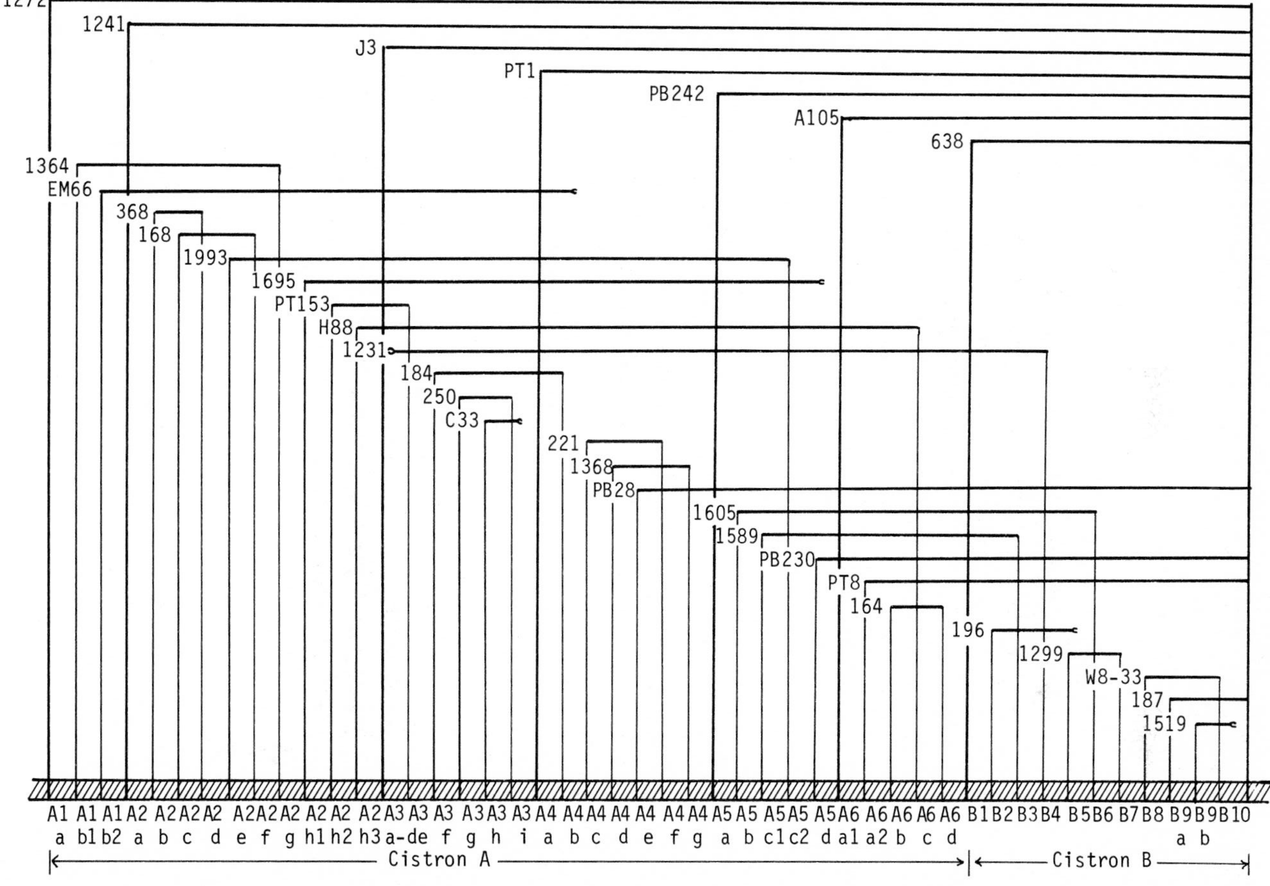

Figure 16b. The mapping of mutations in the rII locus of the bacteriophage T4 [3,55]. The deletion map of the locus, consisting of 47 segments (the right ends of five deletions which did not participate in determining the segments are shown with cavities).

lating influence of his investigations on the work reported in [67] and [80] is unquestionable.

Benzer also carried out a functional analysis of the rII locus, allowing him to fill the classical operational definition of a gene with molecular content.

It turned out that the complementation test divided the collection of rII mutations into two groups, within which mutations are noncomplementary, while any two mutations from different groups are complementary. The groups were compactly projected onto Benzer's map corresponding to the sections he designated as *cistrons*. This term comes from Benzer's modification of the complementation test to the so-called *cis-trans test* (details of his description can be found in [30,43,48]). In Fig. 16b cistron A occupies the left six segments of the "top layer", while cistron B occupies the remaining segment. Thus, according to Benzer, a cistron is a functional unit, having, at the same time, a complex mutational and recombinational structure. By means of some indirect data Benzer was able to closely estimate the physical dimensions of the basic components of his map. It turns out that the mutational and recombinational granularity of a gene ranges down to individual nucleotides, while the functional unit, the cistron, may consist of hundreds or even thousands of such nucleotides. The work of Benzer is one of the most fundamental in the growth of concepts about the genetic systems we described in Section 1.

We now investigate the possibility of using deletions in complementation mapping. We already remarked in Section 1.3 that recombination maps are collinearly recombinational, i.e. they reflect the structure of a locus in case there are mutations, among those being studied, that involve several cistrons at once. Deletions are among the more simple multi-cistron mutations, with respect to their functional manifestation: with them, a loss of function is directly connected with an insufficiency of the corresponding genetic material. The complementation matrix of a collection of mutations, among which are multi-cistron deletions, can be ordered by the methods described above.

As an example we consider the structure of the operon which controls the synthesis of the amino acid histidine in *Salmonella* bacteria. This is one of the most complex operons described up to the present. On its genetic map more than a thousand different mutations have been localized, which produce an insufficiency of histidine in the bacteria [71,81].

The histidine operon numbers nine cistrons, controlling 10 successive stages in the synthesis of the amino acid from biochemical precursors (Fig. 17,a). Benzer's methods described above were used to construct the map (Fig. 17) for preliminary distribution of mutational defects. Moreover, biochemical and complementation data were also used. Thus, the position of cistron H was determined at the beginning, by means of only the complementation matrix [48,81].

The first summaries of complementation data allowed the complete elucidation of the intercistronic topography of the operon by means of deletions. Thus, up to 1964, deletions at the supercistron level had formed the following complementation matrix:

	1	2	3	4	5	6	7	8	9	10	11	12	13	14	15	16
1	1	1	1	1	1											
2	1	1	1	1	1	1	1	1								
3	1	1	1	1	1	1	1	1	1	1	1	1				
4	1	1	1	1	1	1	1	1	1	1	1	1	1	1	1	
5	1	1	1	1	1	1	1	1								
6		1	1	1	1	1	1	1								
7		1	1	1	1	1	1	1	1	1						
8		1	1	1	1	1	1	1	1	1	1	1	1	1	1	
9			1	1			1	1	1	1						
10			1	1			1	1	1	1	1					
11			1	1				1	1	1	1	1				
12			1	1				1			1	1				
13				1				1					1	1		
14				1				1					1	1	1	1
15				1				1					1	1	1	1
16														1	1	1

The matrix is represented in quasidiagonal form. By means of the algorithm of Section 3.1 it is easy to conclude that there are only seven complons in this matrix, one of which (H) is ficticious. However, the linear order of the complons, determined by extended deletions, is collinear with the order of the associated cistrons on the genetic map (Fig. 17,b). The bounds of the deletions are such that the neighboring cistrons E, I and C, D are not different on the complementation map; they are represented by single complons. To separate them it is sufficient to include in the matrix complementation interactions of the deletions with point mutations. If each cistron is represented in the matrix by a point mutation, the corresponding map contains only **real complons**.

Also not represented in the matrix are many cases of intracistronic complementation discovered in the cistrons B, F and E. Their identification was accomplished biochemically. The inclusion of intracistronic

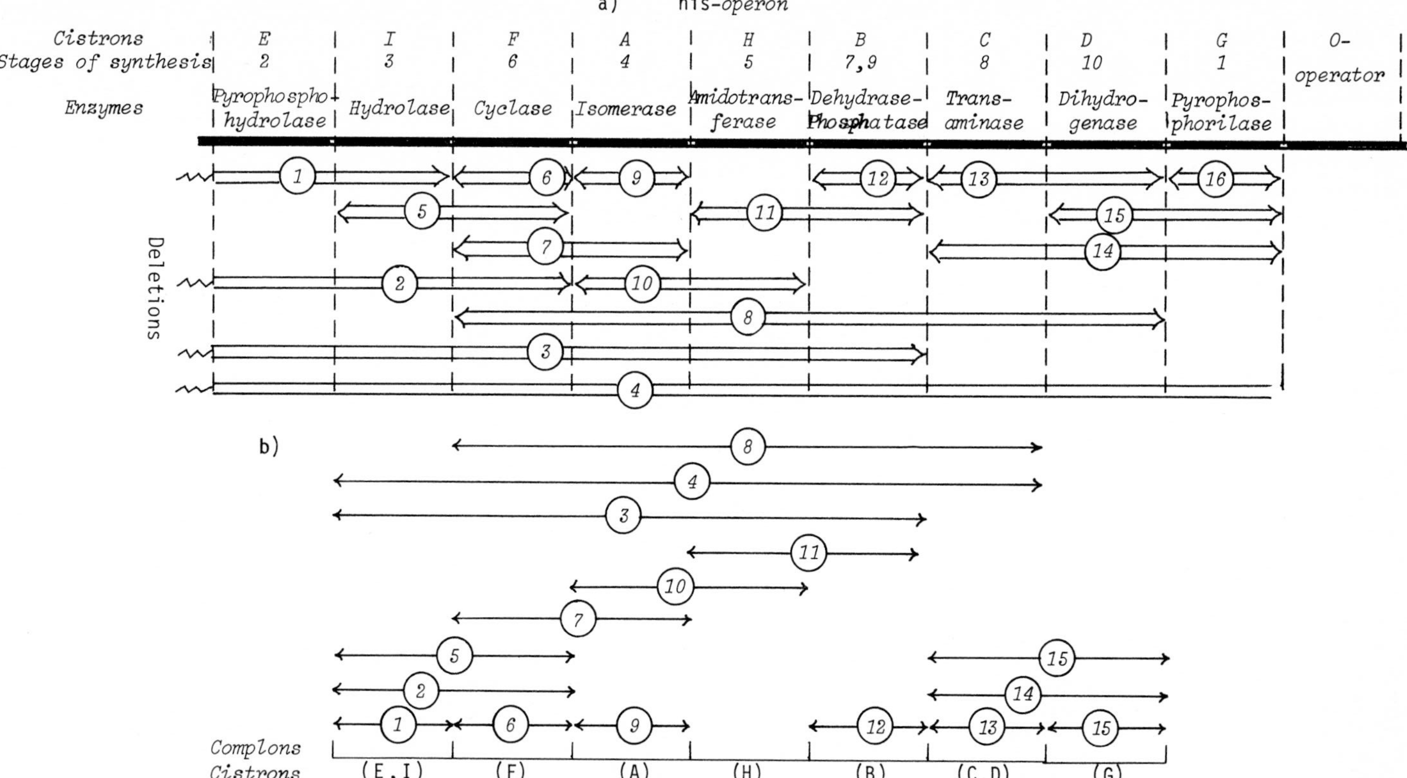

Figure 17. The his-operon map of the bacteria *Salmonella typhimurium*: a) At the top of the map are shown the separate structural cistrons of the operon and the names of the enzymes (O-operator) coded for by them. Below them are several deletions used by Loper et. al. [81] to order the cistrons in the operon (the bounds of the deletions in the cistrons are not shown); b) The complementation map constructed according to the matrix of complementation reactions of the deletions with one another.

complementation in the analysis leads to the impossibility of a linear ordering of the complons (see Section 2.1). However, in this analysis characteristic "concentrations" arise, deviations from linearity, which help determine the cistron groups.

Besides deletions, another class of extensive intercistronic mutations has been successfully applied to the complementation mapping of transcription units in bacteria and viruses. These are the *polar* mutations.

To describe their manifestations it is necessary to briefly dwell on the features of the transcription and translation processes in bacteria. Translation begins before transcription has even been completed. Ribosomes seat themselves on the initial codon of an already-transcribed mRNA fragment and perform the translation process by moving from one codon to the next. As a result, several ribosomes will be seated on an mRNA fragment, each with its partially formed polypeptide chain. On reaching the terminator, the ribosomes release the mRNA with the finished protein molecules.

Suppose that in some cistron a mutation of the type "codon→nonsense" occurs. On reaching the meaningless codon the ribosome, as a rule, cannot proceed because of the absence of the corresponding tRNA. Meanwhile, transcription continues, so that a section of the cistron is formed which is not occupied by ribosomes. Such "free" sections are subject to the action of nuclease enzymes which destroy them as well as the entire remaining chains [30,31]. Therefore, mutations of the "codon→nonsense" type lead to the possible loss of translation of following cistrons.

Thus, although the mutation arises in one cistron, its negative effect is felt in all succeeding ones. Such mutations are said to be *polar*.

We emphasize that polar mutations are possible only in scriptons, where the units of translation (cistrons) are joined at the mRNA level. Besides nonsense mutations, polar effects are frequently associated with the so-called *frameshift* mutations, which result from the loss or insertion of nucleotides. Such mutations change the content (meaning) of all succeeding triplets. Because of this there is a high probability that a meaningless codon will be encountered in the subsequent text.

The nearer the nonsense replacement occurs to the beginning of transcription, the greater will be the extent of the polar defect. A diagnostic feature of all polar mutations is the inactivation of the last structural cistron in the scripton. This is not necessarily so for deletions, although in Benzer's work (see Fig. 16) seven deletions at the top level did have this feature.

Polar mutations permit the easy determination of the linear order of the cistrons within a scripton.

Thus, for example, if among the initial mutations there are only point mutations and polar mutations (Table 2 and Fig. 18), then clearly the point mutations characterize the complons of the map (the cistrons), it being unnecessary to use a general algorithm to order them in this case (page 55). To find the order of the cistrons it is sufficient to find a satisfactory ordering of the polar mutations, according to the numbers of the cistrons (point mutations) they cover.

This method of mapping has been applied to a whole series of bacterial operons: lactose, tryptophan, galactose and others.

An analogous map can be observed for the locus his-3 in the fungus *Neurospora* [54]. In the opinion of some geneticists this is one example supporting the existence of operons in organisms of more complex structure than bacteria. Shown schematically in Fig. 19 are the complementation and recombination maps of this gene. The recom-

Table 2

A complementation matrix †

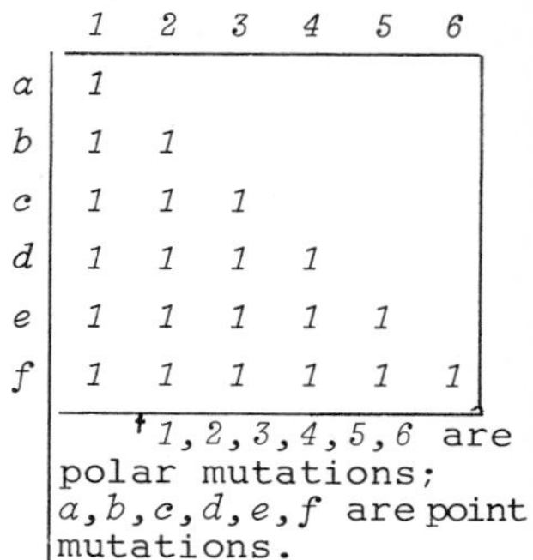

	1	2	3	4	5	6
a	1					
b	1	1				
c	1	1	1			
d	1	1	1	1		
e	1	1	1	1	1	
f	1	1	1	1	1	1

† 1,2,3,4,5,6 are polar mutations; *a,b,c,d,e,f* are point mutations.

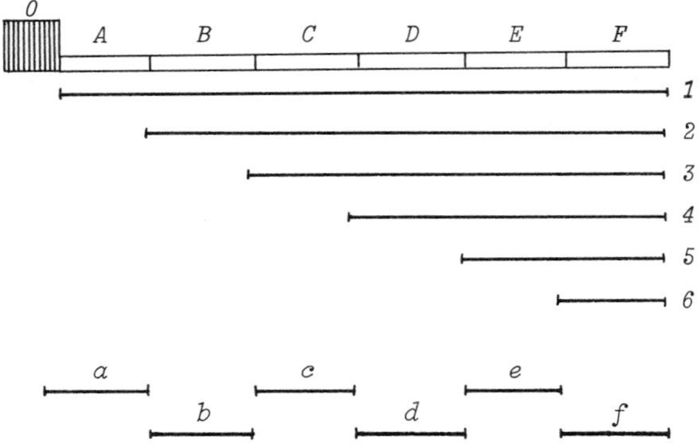

Figure 18. The complementation map of the operon corresponding to the complementation matrix of Table 2.

bination map is divided into five parts: section 0 (possibly an operator) and four structural cistrons (A, B, C and D), which code for corresponding enzymes, three of which are already known. The complementation map contains five polarly complementary mutations, probably due to nonsense replacements. The locations of the mutations on both maps is evidence of their collinearity (the correspondences in position are shown by the vertical arrows).

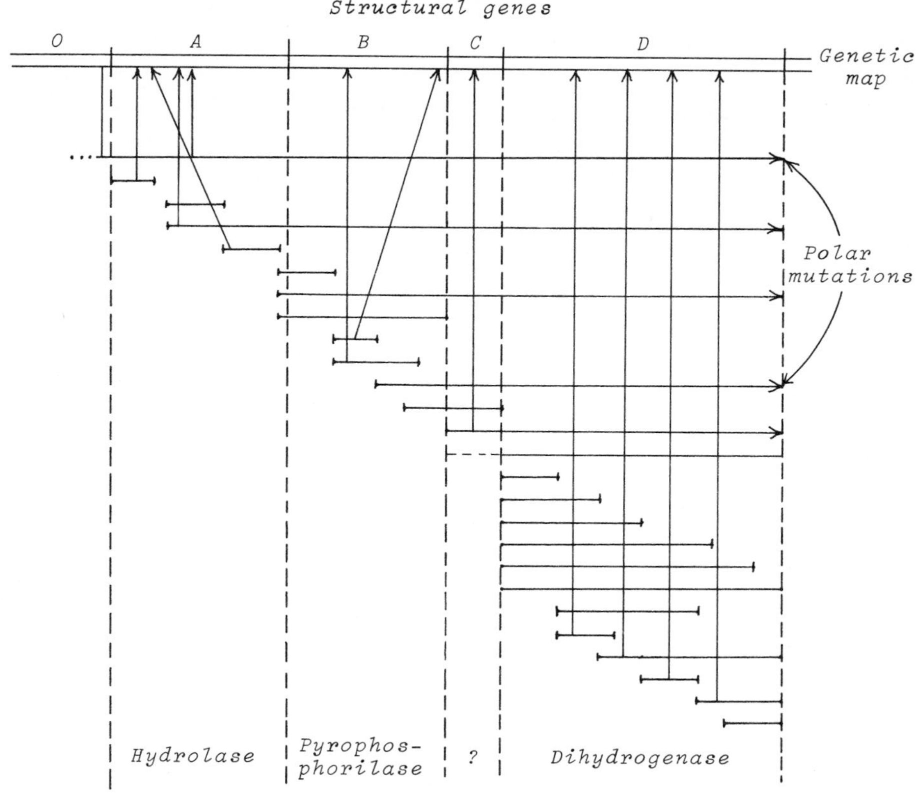

Figure 19. The map of the his-3 locus in *Neurospora crassa* [54]; the complementation map has a sharply dilineated polar character.

2. The complementation maps of multiple mutational defects. There are situations in which it is not possible to construct a linear map by complementation. First of all, the impossibility of constructing unbroken intervals on a linear map (even when all the complons are determined) may be connected with an actual discontinuity of the defects associated with a single mutation. A more exotic type of nonlinearity may involve special chromosome reconstructions called *inversions*, in which the order of the genes in a chromosome section is reversed. Under crosses of two homozygotes with normal and inverted gene positions, progeny heterozygous for the inversion are produced. Suppose, now, that an "inverted" chromosome in one of the parents has sustained a deletion involving two genes which are normally not neighbors (Fig. 20a).

Similar multicistron deletions in this region may arise also in normal (for gene order) parents. Then the progeny, which are heterozygous for inversion and deletion, may provide an intergene complementation matrix (Fig. 20,b) which cannot be represented by a linear map, although all the deletional defects shown in Fig. 20,a are unbroken. This matrix corresponds to a cyclic interval graph (Fig. 20,c).

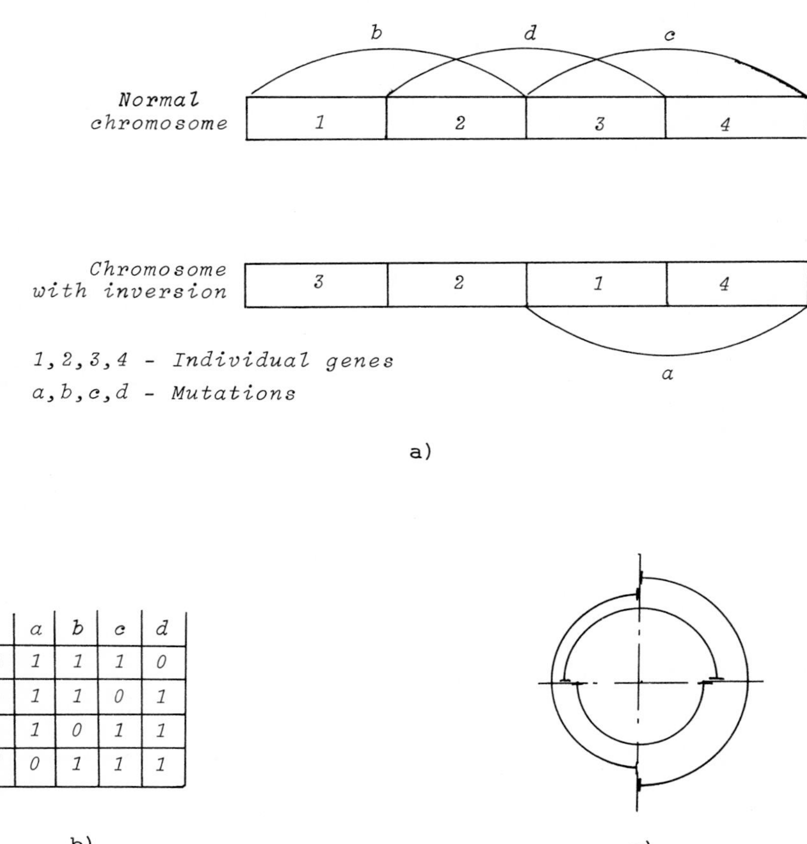

	a	b	c	d
a	1	1	1	0
b	1	1	0	1
c	1	0	1	1
d	0	1	1	1

Figure 20

Naturally, the most simple sources of deviations from linearity are errors in the initial data. Such errors may be connected both with features of the object of study and with low resolution of the genetic technique (not to mention subjective aspects). Therefore, when nonlinearity appears one must first be convinced of the absence of systematic errors of this type.

TABLE 3

Basic complementational "parameters" of recessive lethals, induced in *Drosophila* by foreign DNA's and several viruses.

Series	Mutagen nature	No. of Lethals	Number of complons			
			Total	Real	Fictitious	Isolated
CIV	Cranefly iridescent virus	16	11	4	6	1
DA	DNA of blue-green algae	16	12	4	8	0
C-5	Coxsackie virus, type 5	17	12	2	9	1
C-3	Coxsackie virus, type 3	18	16	3	11	2
IV	Influenza virus (through food)	19	19	2	14	3
DH	DNA of herring	21	20	6	14	0
IVI	Influenza virus (injection)	22	15	2	8	5
DCVD	DNA of CIV + DNAase	22	21	4	16	1
C-1	Coxsackie virus,type 1	24	21	3	16	2
RV	Rous sarcoma virus	25	21	4	8	9
PV	Poliomyelitus virus	27	27	5	21	1
DC	DNA of chick erythrocytes	32	38	4	33	1
DCV	DNA of CIV	41	61	7	53	1
DCT	DNA of calf thymus	66	99	5	90	4

However, the basic, regular source of deviations from linearity in the mapping of a complementation matrix are the cases of interallelic, intracistronic complementation, the molecular events of which do not satisfy the initial premises of linear mapping (see Section 2.1).

We have also concluded the impossibility of linear mapping from the analysis of complementation interactions of mutations induced in *Drosophila* by foreign molecules of DNA, RNA and several viruses [4,37].

These mutations have a recessive lethal action, i.e. in crosses with normal individuals they are not manifest, while the corresponding mutant homozygote flies do not survive. Unlike the usual mutagens (such as X-ray irradiation or chemically active substances), foreign nucleic acids and viruses act on the genetic apparatus of *Drosophila* to some extent directionally: all the mutations elicited by them are localized in a small section of the right arm of one of the chromosomes (the second) [4]. Fourteen different mutagens from this class were examined, to which were matched complementation matrices for the mutations they induced, having dimensions from 16×16 to 66×66 (Table 3).

All the complons were constructed by means of the algorithm given in Section 3.3, but a continuous linear map was not obtained for any of the matrices. Moreover, the continuous variants which we constructed from the maps had necessarily complex topologies: The mutations appeared as G-intervals of the graph G of extremely complex cyclic structures.

What causes such complex structure in the maps? Chromosome rearrangements of the inversion type cannot be the reason. An investigation of the corresponding chromosomes by S. M. Gershenzon showed that inversions and other structural aberrations were encountered very rarely, and even when encountered they were weakly connected with those regions in which complex interval structure was found.

Perhaps the source of nonlinearity is the fact that the mutational defects examined fall within one cistron. The following facts refute this. The chromosome region in which the mutations lie encompasses several dozen (or more) cistrons, and several mutations of particularly complex meshing are located at opposite ends of the region, i.e. they probably belong to different cistrons. Suppose, nevertheless, that some of these mutations belong to the same cistron. Then the cyclic character of the associated map sections implies that we have a situ-

---

These data are the results of many years of experiments on induced mutations, carried out under the direction of Academician S. M. Gershenzon of the Ukrainian Academy of Sciences (the Institute of Molecular Biology and Genetics) [4]. The authors express deep appreciation to him for the opportunity to analyze the data.

ation of the type "defect corrects defect" (the corresponding molecular mechanism is discussed in Chapter 2). But such mutual correction is rare and characterizes specific pairs of defects, while in this case the mutations are complementary to many others.

Thus, there remains the quite simple and natural explanation, that such mutations affect the chromosome in several places at once.

This conclusion is supported by the following regularity (see Table 3). For the smaller matrices the number of complons is slightly less than the number of mutants; for the intermediate-sized matrices these numbers are comparable; and for the larger matrices (the last three mutagens) the number of complons exceeds the number of mutants. An analogous result is obtained when one examines randomly selected mutants from the larger matrices.

In fact, those sections of the chromosome (separated by testing) which are able to mutate individually as well as jointly are the real pre-images of complons. We recall that the complementation map at the level above cistrons, characterizes the mutual disposition of defects associated with mutants. Therefore, if the number of complons equals or exceeds the number of mutants it means that some mutants carry several non-adjacent defects in the chromosome.

How can one nevertheless find a purely linear order for the defects? It is sufficient to merely isolate the groups of mutants representing cyclic map intervals, and carry out direct recombination analysis for them. As a result, those mutants which carry several defects will be determined, and subsequent complementation mapping of these defects (not the mutants) will provide the desired map.

As a basis for this prognosis S. M. Gershenzon undertook, with his colleagues, experiments which completely supported these expectations. The recombination map clearly fixed the desired order of the mutational defects and their multiplicity for specific mutants. It turned out, moreover, that the defect multiplicities obtained are connected with more general and fundamental aspects of the organization of genetic systems. We will discuss this in more detail. What genetic mechanisms have a similar appearance? The distinguishing feature of foreign nucleic acids and viruses, as mutagens, is the specificity of location of their effects in the genetic apparatus of the host. Therefore, understanding of the molecular mechanisms of mutagenic actions is extremely important, since it contributes to the solution of one of the leading problems of genetics — control of the mutational process in multicellular organisms. A number of hypotheses have been expressed that the basis of mutagen action of observed agents lies in their capacity for being taken into the chromosome host-recipient by means of segments homologous to segments of the chromosome [4]. In this process genes near the introduced sgements are destabilized and begin to mutate more frequently.

The multiplicities of induced defects we have revealed provide indirect support for such a mechanism, for the observed mutagens are of the most diverse origins (from RNA of viruses to DNA of the thymus of a calf), and consequently, the segments of homologous identification probably also vary greatly. And in the final analysis, this also leads to multiple defects in a *Drosophila* chromosome: Mutagens of a given type may be present either in a free state, in which case they replicate autonomously, or they may be bound to definite loci of a chromosome, causing mutations in those and in nearby regions. Thus, although local specificity of these mutagens is apparent, it is hardly restricted to one cistron.

Intensive genetic and biochemical investigations in the last ten years have shown that the traditional belief in the invincible stability of a specific genome is nothing more than an antiquated dogma. In the genomes of both higher and lower organisms a large number of varied, mobile genetic elements have been determined, described and partially studied. These are the so-called *insertosomes*, *transposons* (saltatory genes) and other disperse, repeating DNA sequences [69a]. All these elements have a common distinguishing feature: They may change their location in the genome and cause mutational instability in the regions of their intrusion. The frequency of such jumps is several orders of magnitude higher than the average frequency of the usual spontaneous mutation. As a result, multiple defects arise in the genome. Apparently it is such elements that are responsible for the multiple mutagenic effects of foreign DNA and viruses. Moreover, multiplicity of mutational defects itself may be used as an indirect indicator of the presence of mobile elements in the mutational screening of natural populations.

The particular significance of the results described above, in conformity with human populations, adds to the fact that viruses figure here as a mutagenic source (more accurately, the nucleic acids contained in them) [4]. Therefore they should be considered not only as a source of disease, but also as a mutation-causing factor, i.e. as an evolutionary factor. They play this role independently of whether they are infectious. If one considers how widely dispersed viruses are in man's environment and how frequently the two come in contact, the problem of viral mutational action becomes particularly serious in a practical as well as theoretical sense.

3. <u>Complex traits and the loci which control them</u>. It is usual to use recombination analysis (which directly reflects the arrangement of mutational markers) as the means of eliciting the structure of genetic loci. In Section 1 we saw how it is used to localize defects not only on the supercistron level, but also within cistrons. However, recombination analysis of small sections of a chromosome, commensurate with the size of a cistron, requires the examination of tremendous numbers of progeny, since recombinations of nearby markers are so rare. Consequently, investigations similar to those which have been successful for microorganisms are extremely difficult for higher organisms.

The question arises whether the more easily obtainable results of complementation testing can be used in analyzing the structures of such loci. In Section 1 we showed how this could be done when the mutations have a supercistronic character. Although complementation data contain information not about the interactions of actual nucleotide texts, but rather their protein products, i.e. they only indirectly reflect the structure of the genetic material, the success of such endeavors is ensured by the collinearity of DNA, RNA and the primary structure of proteins, functioning as wholes.

However, many phenotypes of higher organisms are variable expression, and binary characteristics of their manifestations in terms of "yes - no" (as was done in the preceding examples) looks very coarse: one can only say that "this thing is and that one is not." In this case the results of complementation testing are not representable as complementation graphs, and the notion of a map is not directly applicable.

Nevertheless, one can try to estimate the similarity of variable expression phenotypes for various mutations quantitatively, and afterwards consider the matrix of indicators of the functional nearness of examined mutations as a "portrait" of the locus. To do this one must step back from the position that structural nearness of defects within the locus being examined determines, to a certain degree, the similarity of their functional manifestations. In other words, the matrix of functional similarity must be reducible to a linear form (Section 4.3) which determines the internal topography of the mutations in the locus (when, of course, the initial data do not include cases of interallelic complementation).

This idea was used in analyzing the functional manifestations of mutations of the *scute* locus in *Drosophila*.

This locus controls the appearance of bristles on the head and thorax of the flies (Fig. 21). Mutations of this locus lead to the loss of growth of some bristles, in which a strict dichotomy is discernible with regard to each specific bristle on a specific fly: the bristle is either normal or non-existent (reduced). Thus, in this case each separate mutant individual is characterized not by one, but by a whole set of binary manifestations — for each of 20 pairs of investigated bristles, i.e. for a 40-element boolean vector. Is each bristle controlled by its own gene? No. Recombination analysis has shown that all these mutations affect the same very small, recombinationally indivisible section of the X chromosome in *Drosophila* — the *scute* locus.

We note that individuals of one genotype carrying the same scute mutation may, generally speaking, have different sets of bristles, so that for a specific mutation it only makes sense to speak of the probabilities of reduction for each bristle. In other words, each mutation of scute is characterized by a 40-element vector of probabilities of reduction in the macrochaeta. The probabilities are estimated according to frequencies of reduction which are quite reliably produced in genetic experiments.

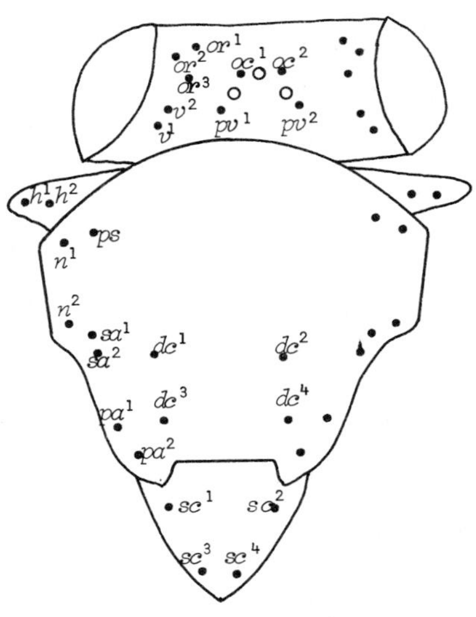

Figure 21. The arrangement of bristles on the body of *Drosophila melanogaster*: *oc—ocellar, pv—post-vertical, or—orbital v—vertical, h—humeral, n—notopleural, ps—presutural, sa—supra-alar, pa— post-alar, dc—dorsocentral, sc—scutellar.*

The *scute* locus was first investigated in the 1920's and 1930's by the Soviet geneticists A. S. Serebrovsky N. P. Dubinin et. al.[5,38]. Their work dealt the first serious blow to the classical concept of an indivisible gene. The gene was divisible, as witness the sectional mutability of the *scute* locus. Under the union of different mutations in a heterozygote, local dominance of norms over reductions was observed with respect to separate bristles. Namely, in each type of heterozygote only those bristles were formed which were in at least one of the original mutations in a homozygote. Of course the correctness of this assertion depends on the frequency of reduction taken as the threshold, above which a manifestation is considered normal, and otherwise, mutant. However, with real data mutant and normal reduction frequencies are quite accurately separable (Fig. 22).

By that time it was already known that different mutations affect compact, unbroken sections of a chromosome (the most simple example is a deletion). Therefore, the next logical step was the premise that for each bristle there is a special section (center) in the *scute* gene that

a)

...	pa^1		n^1		$pv^{1\ 2}$		or^1		$oc^{1\ 2}$		or^2		$sc^{3\ 4}$		$sc^{1\ 2}$	
	l	*r*	*l*	*r*	*l*	*r*	*l*	*r*	*l*	*r*	*l*	*r*	*l*	*r*	*l*	*r*
sc9	52	52	100	100	95	93	100	100	100	100	96	95	100	100	96	96
scD1	44	47	100	100	92	92	100	100	88	88	83	81	99	98	99	99
sc1	5	6	100	100	97	96	100	100	100	100	98	98	84	88	68	60
sc7	1	2	100	100	94	95	100	100	98	99	78	80	99	98	94	96
scD2			100	99	89	88	100	100	95	96	84	85	64	68	55	56
sc6	1	1	93	89	73	79	100	100	99	99	99	99	2	3	1	
sc3B	1	1			34	38	100	100	76	76	74	71	93	92	62	60
sc2B	2	3	1	1	1	2							99	99	97	94
sc260-22					1	2							100	100	100	100
sc5													69	68	18	17
sc8	3	3	2	1	1	1	2	2			5	4	19	19	37	40
scV_2		1	4	4	2	3	6	3			8	9	24	24	23	22

b)

c)

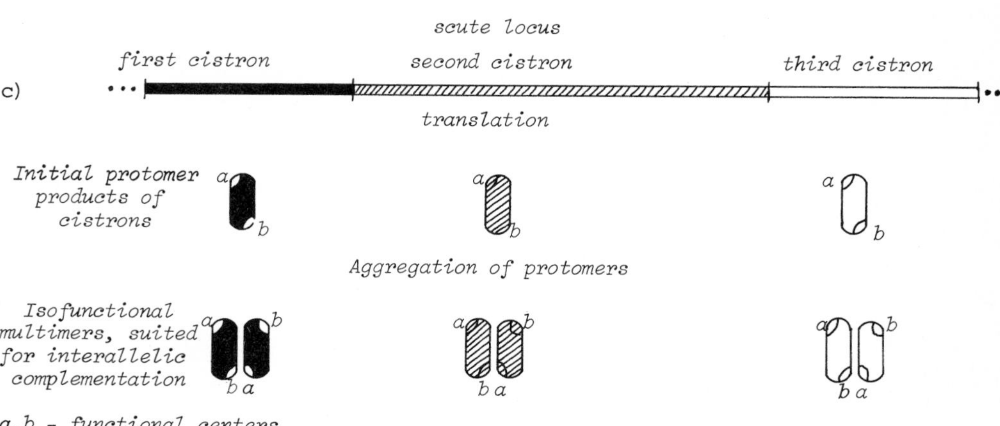

Figure 22. A map of the *scute* locus in males of *Drosophila melanogaster* at 14°C: a) linear map of bristle indices, b) map of mutational defects (the numbers give the percentage reductions in the corresponding bristles), c) a model of the structure and functioning of the locus.

is responsible for its formation, with each mutation affecting neighboring centers. To test this hypothesis A. S. Serebrovsky, N. P. Dubinin et. al. performed a formal procedure which anticipated the method of ordering interval hypergraphs: They found a linear ordering of the bristles for which the indices of those bristles reduced by a given mutation were situated together.

It seemed that the described constructions not only filled a "gap" in the classical theory, but also permitted quite realistic description of the internal topographic structure of the *scute* locus. However, tests of the scute phenomena in various laboratories showed that the phenotypic manifestations of its mutations are very susceptible to changes in temperature and a whole series of other factors [56,68,95]. This contradicted a basic property of genetic material — the stability necessary for faultless transmission of hereditary information from generation to generation.

Since the character of reduction changes with change in temperature it raises doubt whether this phenomenon has a real connection with the structure of a stable locus.

For example, the noted geneticist R. Goldschmidt [68] felt that such a map reflects a sequence of events in the formation of bristles over time, the course of which permits dependence on such factors as temperature. As a result of this criticism work on the genetic analysis of the scute system ceased.

Recently, complementation phenomena in the *scute* locus were investigated anew with dependence on temperature [33,45].

Bristle reduction frequencies for 468 genetic combinations of 12 scute mutations were analyzed. Calculations of the frequencies for each combination were produced for a set of 400-500 flies. Crossing data for $12 \times 11 = 132$ heterozygotes and 24 homozygotes (by twelves separately for male and female, since the trait being investigated is coupled with sex, and its appearance differs accordingly) were examined. In all, 156 combinations of the genotype were examined, each one for three temperatures (14°, 22° and 30° C), constituting a total of 468 experiments.

To estimate the functional nearness of mutations i and j ($i,j = 1,2, \ldots, 12$) we calculate the distance between the 40-element vector of reduction frequencies corresponding to the homozygote for mutation i, and the analogous vector corresponding to the homozygote for mutation j, by the formula

$$f_{ij} = \sum_{s=1}^{40} |r_{is}-r_{js}|,$$

where i, j are the indices of mutations (alleles), and r_{is} is the reduction frequency of the sth bristle for the ith allele.

Thus twelve distance matrices $\|f_{ij}\|_1^{12}$ were obtained (separately for males and females for each of the temperatures).

One of these matrices is presented in Table 4. It is not difficult to convince oneself that this matrix cannot be reduced to linear form. However, with the ordering given, the number of deviations from linearity is not great, as one can distinguish its block structure with the "unaided eye": The first and last sets of five mutations form groups which are not connected with one another, since the intragroup distances are noticeably less than those between groups. A similar picture holds for the remaining five matrices: The same three groups of mutations are discernible, with the "almost linear" forms of the matrices obtained by ordering the alleles differing from this one in only one case, in which three permutations of neighboring mutations within groups occur. This supports the existence of substructure in the *scute* locus corresponding to the three groups obtained. These groups are invariant with respect to changes in temperature, while the proteins (unfortunately, as yet unidentified), which evidently also define the observed phenotype, must significantly change their activity with temperature fluctuations. Consequently, the invariance of these groups is explained by the structural stability of the gene, not by the corresponding proteins. Moreover one can conjecture that the substructures of the locus corresponding to these mutational groups are cistrons, because it is they that code for separate proteins. With a change of temperature the order of the mutations changes only within the groups, which easily explains the changes in protein activity.

We will now trace the extent to which our speculative construction is connected with the real process of bristle formation. Do the invariant orderings of mutations we have obtained correspond to some kind of regularities in bristle reduction? To answer this question consider the table of reduction frequencies for homozygotes (see Fig. 22). Columns (bristle indices) were ordered so that for each mutation (row) bristles with the largest reduction values fell together (i.e. the matrix was row linear), while the ordering of rows was obtained from the preceding distance matrix analysis.

As is evident from Fig. 22, the form of the table obtained is nearly linear (for rows). Consequently the order of bristles also reflects the locus structure.

We see that mutations of the various groups correspond to different types of rows: those of the first group correspond to long rows, those of the third to short rows, and the second group to intermediate-length

TABLE 4

An ordered matrix of the functional differences of scute mutations, constructed according to percentage reductions of bristles in mutant homozygous females at 14° C (the separated blocks are of mutations functionally most similar to one another).

	scD1	sc9	sc7	sc1	scD2	sc6	sc3B	sc28	sc260	sc5	sc8	scV2
scD1	0	2694	3418	4078	5178	6563	7434	11966	11162	13648	15850	16971
sc9	2694	0	2506	3136	3388	5877	5484	10374	10276	11714	13528	14925
sc7	3418	2506	0	4076	3324	6033	5432	10208	9584	11490	13712	14601
sc1	4078	3136	4076	0	2528	2863	4744	11124	11904	10330	12508	13825
scD2	5178	3388	3324	2528	0	2841	3176	9320	10344	8742	10716	12025
sc6	6563	5877	6033	2863	2841	0	5095	11243	12271	9361	10481	11922
sc3B	7434	5484	5432	4744	3176	5095	0	6420	7232	6358	8584	10237
sc28	11966	10374	10208	11124	9320	11243	6420	0	1056	1890	4064	6083
sc260	11162	10276	9584	11904	10344	12271	7232	1056	0	2918	5092	7107
sc5	13648	11714	11490	10330	8749	9361	6358	1890	2918	0	2334	4248
sc8	15850	13528	13712	12508	10716	10481	8584	4064	5092	2334	0	2643
scV2	16971	14925	14601	13825	12025	11922	10237	6083	7107	4248	2643	0

rows.

The columns of the table of Fig. 22 also lend themselves to natural division into three groups. The right-most of these groups is engendered by the short rows of the third mutation group. The exactness of division of the remaining columns into two groups flows from the constitution of analogous tables for different temperatures. In the columns of the left group reduction frequencies change more or less monotonically with change in temperature in all the mutants, without exception. The same regularity holds for columns of the third group, while in the columns of the middle group dependence of reduction on temperature has a complex, irregular character (for more details see[45]).

We not only confirmed the preceding conclusions, but also obtained a number of additional ones. The principle one is that individual mutations can be interpreted in terms of the linear ordering of bristles, so that the bristle groups which have been determined correspond to separate cistrons, resulting from analysis of the matrix of intermutation distances. In this the structure of the table points out the possible polar character of the mutations peculiar to scriptons and operons.

The above considerations allow one to propose the following model of structural organization in the *scute* locus (see Fig. 22).

It is conjectured that the *scute* locus is a gene of operon type, consisting of a small number (3-5) of structural cistrons coding for proteins which fulfill the same function, but in different parts of the fly body, forming different bristles. In accord with the model, the first five mutations affect all three cistrons. The defects corresponding to them may either be the usual deletions of all the cistrons, or begin in the first cistron, affecting all information subsequently read out, as in the case of polar mutations of the nonsense or frameshift type in operons. The last five mutations are point mutations and affect only the final cistron. Several fall in the general picture of mutations sc6 and sc3B. Their nature is unknown, so that their position on the map cannot be accurately determined. In Fig. 22 the bristles controlled by each cistron are evident. In this connection it is interesting that each cistron controls its own compact portion of the fly body (compare Figs. 21 and 22), although there is no correlation between the order of the cistrons in the map and corresponding locations of bristle formation on the body.

How can this model be verified? Direct methods for investigating the structure and functioning of loci like *scute* are still not possible. We therefore revert to the usual method of verification: We draw from the model consequences concerning phenomena directly observable in experiments. If the observations contradict the consequences the model is declared to be groundless, otherwise the model is supported (although indirectly). We will attempt to take advantage of this method, bearing in mind that up to this time we have no material in any way reflective of the behavior of scute mutations in the heterozygous state.

Thanks to the elegant experiments of C. Stern [94] we know that the plan of arrangement of the bristles on the fly body (called prestructure) is not under control of the *scute* locus; other genes are responsible for it. The *scute* locus plays the role of a distinctive "commutator", triggering the bristle formation process in response to a specific factor ("signal") from the prestructure. Our model implies that such a triggering occurs at the post-translation level: From among all the isofunctional proteins synthesized by the *scute* locus some one is activated by a factor specific to it, and afterward the formation process begins. The specific "identification" of prestructure factors with protein products of the *scute* locus means that they possess allosteric properties (we will discuss these in Chapter 2) and therefore they must have quaternary structure.

The analysis of scute mutation behavior in heterozygotes can serve as indirect support for this conclusion. If homo- and heterozygous mutant manifestations are compared as such, before their reduction to the binary values (0,1), then it is found that besides the local dominance of reductions by the norm, in heterozygotes reduction on some bristles may be either equal, lower (complementation), or higher (anticomplementation), than in each of the mutant homozygotes. But as we will see in Chapter 2, such complementation effects are inherent only for proteins with quaternary structure. In passing, we remark that in the model of an operon with a complex structure in its operative section, suggested by V. A. Ratner for the *scute* locus [29,31,33], such effects are theoretically unexplainable.

We now consider what kind of picture one can expect for functional nearness of mutations, if nearness is estimated for heterozygotes under the assumption that our operon-like model holds.

We fix some mutation i and consider all 11 heterozygotes of type (i,j), where $i \neq j$. We will call mutation i the reference for this collection of heterozygotes. What does our model say about the qualitative features of functional nearness in heterozygotes?

First consider the case when the reference mutation is from the first group. These mutations affect (inactivate) all three cistrons, so that bristle formation is determined by the second allele in the heterozygote. Thus, the nearness of heterozygotes (i,j) and (i,k) is determined by the nearness of the second components j and k. This means

TABLE 5

The ordered matrix of functional differences of scute mutations, constructed according to percentage reductions of bristles in mutant heterozygotes: sc1 X sci at 14° C (blocks of mutants which are functionally most similar are the same as for the homozygous case.

	scD1	sc9	sc7	sc1	scD2	sc6	sc3B	sc28	sc260	sc5	scV2	sc8
scD1	0	1719	1839	2058	2867	3582	4717	7836	7447	9147	5750	8649
sc9	1719	0	1930	1437	2282	3253	4178	7557	7482	8688	5619	8290
sc7	1839	1930	0	1283	1676	3221	3436	6611	6206	7892	4755	7410
sc1	2058	1437	1283	0	1559	2518	3375	7288	7211	7861	4788	7357
scD2	2867	2282	1676	1559	0	1753	2176	6511	6416	6460	4931	6148
sc6	3582	3253	3221	2518	1753	0	2537	7480	7385	6271	5700	5921
sc3B	4717	4178	3436	3375	2176	2537	0	5301	5704	4558	5435	4672
sc28	7836	7557	6611	7288	6511	7480	5301	0	563	1379	2986	1657
sc260	7447	7482	6206	7211	6416	7385	5704	563	0	1770	2891	1886
sc5	9147	8688	7892	7861	6460	6271	4558	1379	1770	0	3807	664
scV2	5750	5619	4755	4788	4931	5700	5435	2986	2891	3807	0	3323
sc8	8649	8290	7410	7357	6148	5921	4672	1657	1886	664	3323	0

TABLE 6

The ordered matrix of functional differences of scute mutations, constructed according to percentage reductions of bristles in mutant heterozygotes: sc8 × sci at 14° C (the matrix is not reducible to linear form, mutants cannot be separated into blocks).

	scD1	sc9	sc7	sc1	scD2	sc6	sc3B	sc28	sc260	sc5	scV2	sc8
scD1	0	380	385	304	368	439	339	280	388	332	509	1364
sc9	380	0	369	362	432	537	549	370	562	446	535	1464
sc7	385	369	0	483	377	600	648	521	665	543	672	1613
sc1	304	362	483	0	448	337	271	288	316	214	335	1288
scD2	368	432	377	448	0	633	669	562	688	608	695	1648
sc6	439	537	600	337	633	0	174	323	261	221	368	1229
sc3B	339	549	648	271	669	174	0	229	135	129	348	1137
sc28	280	370	521	288	562	323	229	0	248	168	403	1216
sc260	388	562	665	316	688	261	135	248	0	164	319	1152
sc5	332	446	543	214	608	221	129	168	164	0	297	1168
scV2	509	535	672	335	695	368	348	403	319	297	0	1209
sc8	1364	1464	1613	1288	1648	1229	1137	1216	1152	1168	1209	0

that the matrix of functional distances must have a near-linear form for the same ordering of mutations as in the homozygous case.

If the reference mutation is from the third group then elements of the matrix of functional distances of second components must be small, and the matrix must be far from linear and from a block structured form. The fact is that in this case the reference mutation does not affect the first cistrons, so that in all combinations of heterozygotes the qualitative composition of functioning proteins is the same (although their concentrations may vary, depending on how many cistrons affect the second component).

The data at hand completely confirm these conclusions. In Tables 5 and 6 both cases are shown. The form of the matrix for the reference mutant sc1 from the first group is the same as the one for homozygotes (nearly linear, with a similar ordering of mutations, the same blocks of mutations near to one another).

For the matrix of functional distances of heterozygotes for the reference mutation sc8, from the third group, our prediction also holds: matrix elements are small, the matrix itself is far from linear, the mutations are not divided into blocks. The same is true for other mutations (considered as reference mutations) from the first and third groups, respectively.

This provides the desired experimental confirmation. From this model a number of more subtle conclusions also follow, concerning the specific functioning of protein-multimers (the products of structured cistrons) in corresponding sections of the fly body, which also are amenable to experimental genetic testing (and sharpening of the model).

Thus, the results obtained indicate that the initial linearity of structure of the genetic material is nevertheless sufficiently stable to be manifest even after a long chain of intermediary influences from various ontogenetic factors.

Of course final confirmation of this model must await direct biophysical investigation of this system.*)

---

*) Very recently the scute-region of the X chromosome has been cloned and mapped by the "direct" restriction method 55b . The comparison of these new biochemical results with ours is premature because the number of scute mutations investigated at DNA level is too small. It should be noted, however, that DNA localization of mutations sc1, scD1, scD2, sc6 and sc3B *correlates with our cluster-polar map*.

Chapter 2. Graphs in the Analysis of Gene Semantics

§1. Interallelic complementation and the functioning of protein multimers.

1. Interallelic complementation. The complementation test, unlike the recombination test, has an initial functional nature: in it the differences in mutant functions are considered. However, at the inter-cistronic level of complementation events (by the analysis of scriptons and operons (1.5.1)) this test permits the characterization not only of the functioning of separate cistrons, but also the structural topography of a system as a whole. The molecular-genetic basis of this is, on the one hand, the independence of cistron translation and the subsequent functioning of the proteins which correspond to it, and on the other hand, the presence of polar mutations which, having appeared in a single cistron, block the action of all subsequent ones. Because of this, polar mutations lead to the violation of the classical rule that mutations of different cistrons are complementary to one another.

In this chapter we discuss a different kind of violation of the rule, intracistronic complementation, in which the complementary mutations are within the bounds of a single cistron. This phenomenon, called *interallelic complementation* was discovered at mid-century. Later, it was established, that interallelic complementation is characteristic for those cistrons which code for proteins having quaternary structure, i.e. which are complex aggregates of separate polypeptide chains, produced by the cistron [43,57]. Thus the test for interallelic complementation is a genetic method of finding out whether a protein has quaternary structure. If there is no quaternary structure then complementation within the corresponding cistron is ruled out.

Here, the objects to be crossed are mutant homozygous individuals. The offspring of such crosses then are heterozygous for the given cistron: one defect of the cistron has paternal origin and the other maternal origin. Each of these cistrons repeatedly synthesizes the corresponding form of the modified polypeptide chain. It is assumed that the aggregation of individual instances of these chains (monomers), during the formation of quaternary (multimer) structure, takes place randomly.

If a multimer is formed by the union of identically altered monomers (of one type or another), then the proteins obtained are identical to the parental ones. However, the aggregation of differently altered monomers gives rise to a new hybrid form of the protein multimer which is absent in the parents. It is known that such hybrid proteins are responsible for new functional properties of the heterozygous offstring, which is the manifestation of interallelic complementation.

Usually the phenomenon of interallelic complementation is studied in microorganisms (viruses, bacteria, fungi etc.). Initial mutant parents unable to grow in a well-defined medium due to insufficient activity of the corresponding mutant protein are used. If, as a result of crossing of these mutants, growth is observed in the descendant colonies (complementation), then this is clearly due to a hybrid fraction of the protein multimer, since it alone distinguishes the offspring from the parents. This means that the hybrid protein multimer, consisting of homologous but differently altered monomers, displays greater activity than that found in the parents (which contain identically defective subunits).

These phenomenological peculiarities point out the fact that complementary mutations of a single cistron must be sufficiently profound to inactivate the protein in a homozygous situation, but at the same time, not too serious, so that renewed function (even if only partial) is ensured in the heterozygote [35,40,88].

Thus, to understand the molecular mechanisms of such renewals-of-function it is necessary to discuss the principles of the construction and functioning of protein multimers.

2. The fundamental principles of organization of protein multimers. Proteins are the most complex macromolecules of cells. Direct physico-chemical investigations of their structures are very laborious, and for proteins occuring in the cell in insignificant concentrations (and there are many such) it is simply impossible, with current methods.

To the present time, detailed atomic structure has been determined for only a few proteins: globins, lysosomes, dehydrogenase and others [49,53,86,87].

From a functional point of view proteins are constructed as follows. On the tertiary and quaternary levels proteins form special spatial structures called *functional centers*, each of which fulfills a definite operation, connected with overall protein function. Among the kinds of functional centers in proteins are: *catalytic centers*, which directly process substrates; *contact centers*, those parts of a monomer at which it comes in contact with other monomers of the multimer; *allo-*

steric centers, through which the activity of the protein is regulated by external, low-molecular cell substances; and a number of others.

As an example we consider the functional organization of hemoglobin, a protein which has been studied rather thoroughly in recent years. In particular, in the last five years the structure of the gene which codes for hemoglobin in various species has been completely deciphered by means of DNA sequencing techniques (see below, page 159 and Fig. 43'). *Hemoglobin*, a transport protein, participates in a most important function in animals, oxygen respiration. The basic results concerning the molecular structure of this protein were obtained by M. Perutz and coworkers as the result of many years of investigation [87]. Hemoglobin is of interest in this book not only as a thoroughly studied molecular object for demonstrating the principles of semantic analysis of genetic texts, in this chapter, but also as a specific object for genetic discussion in Chapter 3.

Normally a mammalian hemoglobin molecule appears as a *tetramer*, consisting of four monomer subunits of two types, designated α and β. The conventional arrangement of the tetramer is shown in Fig. 23, including the contact points between separate subunits.

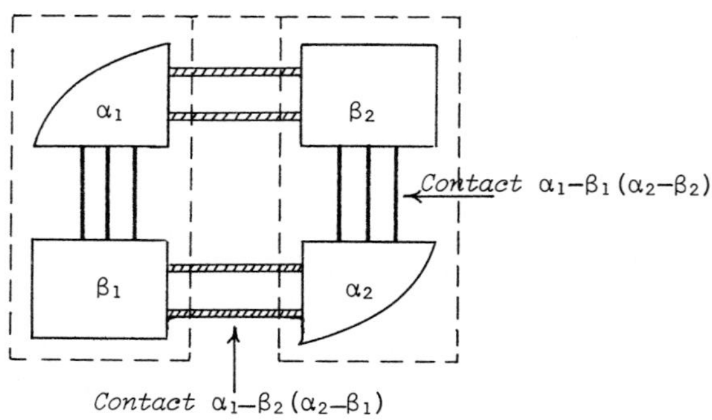

Figure 23

Each chain of the tetramer has a complex spatial packing, and connects to two chains of the opposite type, as shown in Fig. 23. Under chemical dissociation the molecule at first separates into the two dimers $\alpha_1\beta_1$ and $\alpha_2\beta_2$ outlined in Fig. 23 by dashed lines. The centers of mutual identification (contact) of the subunits are designated α_1-β_1 and α_2-β_2 within the dimers, and α_1-β_2 and α_2-β_1 between the dimers. Although the α and β subunits are coded by different cistrons, their structures are very similar, which clearly points to the common character of their origin in the evolutionary process.

Protein multimers consisting of identical, or homologous (as in hemoglobin) chains are symmetrically structured. The hemoglobin tetramer has three axes of symmetry, with the contacts α_1-β_1 and α_1-β_2 being realized near these axes [1,49,86,87].

The most important semantic part of any protein molecule is its active center. In the transport molecule, hemoglobin, the organization of the active heme-specific center includes, aside from the set of amino acids, a pigment molecule called *heme* (containing an iron ion), which binds a molecule of oxygen. Thus the heme-specific center recognizes as a substrate an oxygen molecule.

It is curious that the amino-acid segment of the heme-specific center is also involved in one of the contact centers (in the chain α_1 it is the contact center α_1-β_2). This plays a decisive role in the functions of binding and releasing oxygen molecules, performed by the hemoglobin tetramer. An oxygen molecule bound to one chain changes the conformation of its heme-specific cavity, and consequently, of the contact center between the dimers, thus transmitting the change to the other subunit. As a result the probability of the other subunit taking up the next oxygen molecule is increased by two orders of magnitude. Thus, because of the quaternary nature of its structure, the binding of even one oxygen molecule automatically leads to a sharp acceleration in the absorbtion of subsequent molecules (by a restructuring of the recognizing apparatus).

Interactions of this kind, between different centers of proteins, are said to be *allosteric*. Clearly, the appearance of quaternary structure in proteins is due primarily to the advantages connected with allosteric interactions, which provide cooperative functional effects [16,49].

The role of the contact centers α_1-β_2 and α_2-β_1 in hemoglobin is thus of particular importance, since they also facilitate allosteric interactions. In a number of vertebrates, notably mammals, an important allosteric function is carried out by a small center situated in the β_1-β_2 contact zone (consisting of only four amino acids), which binds to 2,3-diphosphoglycerate. The bound organic phosphate is an intermediate product of carbohydrate decomposition (glycolysis); and when bound, it lowers the affinity of hemoglobin for oxygen. In other words, by this means the process of blood saturation by oxygen is regulated, depending on the intensity of the metabolic processes of the body. However, in general, contact centers and allosteric centers can be spatially separate, as they are in a number of enzymes which have been studied.

Thus the functioning of a single hemoglobin chain, for example α_1, may be represented in the following way. A newly-translated α-chain includes heme in its tertiary structure (after identifying its heme-specific cavity), and is then aggregated with a β-chain to form a dimer (by means of the contact $\alpha_1-\beta_1$), which then combines with another dimer (contact $\alpha_1-\beta_2$). The binding of an oxygen molecule by a chain is possible in one of two states, determined by the contact $\alpha_1-\beta_2$, by which the chain "knows" whether there is an oxygen molecule on the other dimer. In tissues with a lowered concentration of oxygen the subunit gives up its oxygen molecule.

Analyzing the details of the structure and interaction of functional centers, V. A. Ratner reached the conclusion that the cistron instructions which code for the α-chain of hemoglobin can be represented in the following form [31]:

If heme is recognized then
If the center $\alpha_1(\beta_1)$ is recognized then
If the center $\alpha_1(\beta_2)$ is recognized then
If the center $\alpha_1(\beta_2)$ is not restructured then
Bind a molecule of O_2 with reduced affinity
Otherwise
Bind a molecule of O_2 with increased affinity;
If the concentration of oxygen is below the threshhold level then
Release a molecule of O_2.

This phrase illustrates very well a basic property of the genetic language: the imperative form of the protein sentences, which consist of "cases" or conditions for the execution of various commands. In this there is a strong similarity between genetic information and the "machine" texts of algorithmic languages.

As we see, the semantic analysis of cistrons requires knowledge of the functional centers of the corresponding proteins, including knowledge of the molecular character of their functional interactions. Hemoglobin is practically the only example of a protein for which researchers have obtained a more or less detailed knowledge of this kind. Exact physico-chemical analysis of proteins by modern methods is so laborious and cumbersome that it requires years and years to decipher the semantics.

This is why attempts to determine the semantic properties of genes by some other method are so attractive. The analysis of interallelic (intracistronic) complementation interactions of mutations seems particularly promising for this approach, since it is complementation that is connected with the meaning of genetic information as manifest in protein functions.

3. <u>The molecular mechanisms of interallelic complementation</u>. It is clear that, at the intracistronic level, the possibility of complementation of mutations depends on the constitution of the functional centers in defective monomers, aggregated together into multimers.

Until recently the model generally used was that of F. Crick and L. Orgel [57]. In their model the initial molecular events of interallelic complementation take place in the contact centers of the hybrid protein multimer, since it is there that the defects of the various subunits are spatially near together and can have influences on one another (see also [40,43]).

Mutational replacements of separate amino acids (and it is only these mutations we are concerned with) lead, at the level of tertiary structures, to the alteration of the conformation of the polypeptide chain. Generally speaking the sizes of these regions differ, depending on the character and the position of the replacement, since the tertiary configuration of a protein is completely determined by its primary structure.

The fact that certain proteins are only active in heterozygotes, and not in homozygotes, leads one to notions of the "norm correcting for the defect" and to principles of covering or non-covering of defects similar to those for intercistronic complementation (but on the protein level). According to Crick and Orgel, if the hybrid multimer is not active, then it is as if the altered zones of conformation in the mutants are partially overlapping in the region of contact of the subunits. On the other hand, even a partial renewal of activity corresponds to non-overlap of the altered zones. This aspect justifies the importance of mapping data about interallelic complementation by the methods of interval graphs (Chap. 1): If the mutations alter homologous contacting portions of subunits (as is always the case with mutant homozygotes), then they are non-complementary; otherwise the normal part "corrects" the defect homologous to it.

It is important to emphasize that such a mechanism for homologous correction of a defect by the norm works near the axis of symmetry of the multimer. Therefore, according to Crick and Orgel, the mutual disposition of the sections of the homologous contacting subunits in the multimer must be reflected in the complementation map.

However in recent years there has accumulated a considerable amount of data, both of genetic and purely biochemical kinds, which does not agree with this model.

We have already mentioned the fact that the functional centers of a protein are not represented, in the protein's primary structure, by compact fragments. Similarly a mutational deformation of the centers is also not compact, but "spread" over the whole amino acid sequence.

But then the first postulate of complementation mapping has already been violated, since such a mutation cannot be represented by an interval on a line. Therefore, in particular, it is too much to expect any appreciable coordination of the positions of mutational defects in complementation (functional) and recombination (structural) gene maps. Moreover, considering the latest data concerning the multi-domain organization of a number of proteins (see for example the immunoglobulins on page 7), in which domains are sequentially projected onto the primary protein structure (and, correspondingly, onto the recombination map), one can expect a correlation between the genetic and complementation maps to within the occurrence of domains. In accord with the hypothesis that domains are coded for by separate exons, the situation here is completely analogous with intercistronic complementation, ex-

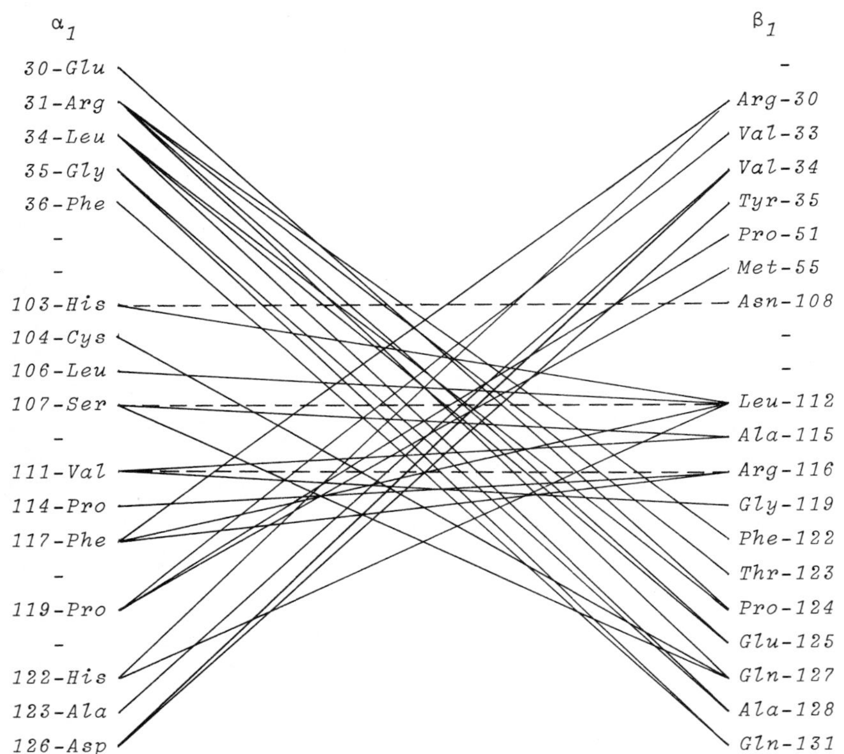

Figure 24. The structure of the contact centers α_1-β_1 in hemoglobin: Opposite each position in the α-chain is its homologous position in the β-chain; connections between amino acids in homologous positions are designated by dashed lines; those between non-homologous positions, by solid lines.

cept that exons replace cistrons.

The notion of mutually homologous correction of protein defects has also been shown to be faulty, since it turns out that the subunits of a multimer are situated in "antiparallel" positions [53,87], so that as a rule the homologous parts of different subunits are distant from one another, and not in contact (Figs. 24 and 25). Consequently, there is simply no place that the mechanism of Crick and Orgel works.

It is known that the pnenomenon of intracistronic complementation is "tied to" the basic semantic substructures of proteins: the func-

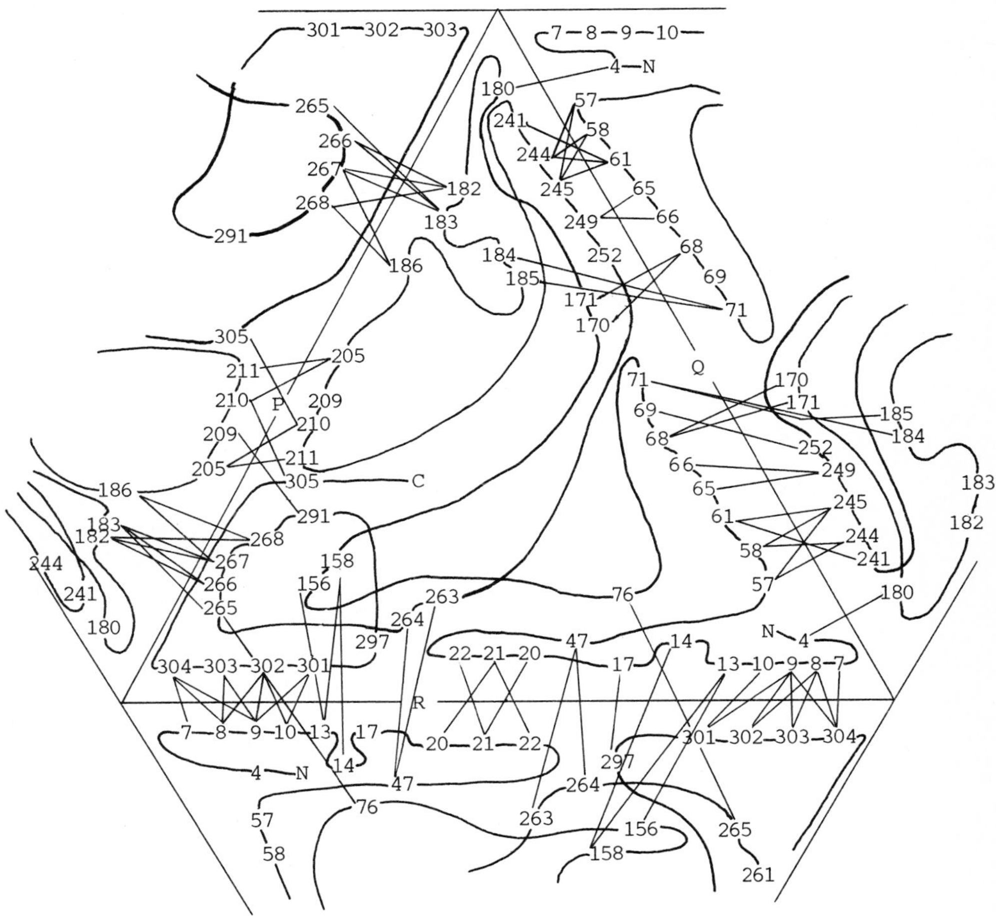

<u>Figure 25</u>. The structure of the contact centers of subunits in an isoenzyme lactatedehydrogenase of a shark [53]: P, Q and R are the axes of symmetry of the multimer; the numbers designate the order of the amino acids in the primary structure; lines join the amino acids which are in contact.

tional centers. Therefore any model of the mechanism of interallelic complementation must also be based on the organizational features of these centers.

In a paper by Ratner, Rodin and Shenderov [35] a model is examined which satisfactorily explains the fundamental known facts (see also [43,59,88]). The basis of this model is the fact that each subunit of a multimer contains a complete set of its functional centers, so that in a multimer each type of center is represented by several copies, depending on the number of subunits. Each mutation is able to block one or more centers of separate chains of the multimer. We will consider that a multimer is able to fulfill its function if it contains even one correctly-working copy of a functional center of each type, i.e. in the hybrid multimer there are no common affected centers (complementarity). If in a hybrid multimer identical centers are affected then these centers are blocked in all the subunits of the multimer, and as a result it is inactivated (non-complementarity).

Contact centers occupy a critical position. Effects on contact centers frequently lead to the "break down" of the multimer, independently of whether the mutations are homozygous or heterozygous. However if both mutations affect the same contact center, then it is possible that they are complementary, i.e. mutually-correcting. This is connected with the fact that in the contact zone the amino acids of different subunits interact directly (physico-chemically), and this ensures fulfillment of the contact function. The normal character of interactions in a contact zone may be preserved by the "corrective" influences of the mutually-contacting defects, which leads to complementation. Such mutual corrections of the amino acid replacements participating directly in contact centers must be very specific, and consequently, infrequent. This is because of the physico-chemical individuality of connections between amino acids in the contact zones of subunits. It is another matter, if the mutations affect the contact centers indirectly, by means of their nearby environments, so that the amino acids participating directly in the centers are preserved and only the conformation of the centers is changed. It is easy to understand in this case that for mutual correction of defects it is not necessary to have stringent, mutual, physico-chemical specificity of replacements but only that the stereometry of that part of the multimer be maintained [35,88].

It may appear that the mechanism we have described reiterates the model of Crick and Orgel. This is not so, however: With Crick and Orgel "the good corrects the bad", while here "the bad corrects the good." this is possible because the subunit contacts are nonhomologous by their very nature.

This model can be called a *mosaic* model, since complementation effects are defined by a distinctive mosaic of normal copies of the functional centers of a protein multimer.

In realizing the function of a protein the centers interact with one another, as we discussed in the example of hemoglobin. This mutual interaction is reflected in the behavior of mutations which affect more than one center, in particular through their complementation reactions. If the mosaic model is correct, then mutations which are similar in their complementation manifestations must correspond to the same functional center. In other words, the structure of the basic complementation relations of a mutant system must correspond with the structure of the interactions of the functional centers, and not with the "mythical" structure of the arrangement of mutational defects. Actually, for a number of well-studied proteins (of the hemoglobin type) one notices that mutations which effect the same center are similar in their functional manifestations; running all the way from local molecular changes of protein function to the pathology at the level of the whole organism [30,46,70,86].

It is clear from the foregoing that the analysis of the structure of intracistronic complementation matrices presupposes the identification of the functional centers and the structure of their interactions.

The formulation and investigation of mathematical problems of structural analysis in relation to the mechanism described above, is given in the following section. It must be mentioned that, as will become clear, these formulations are applicable in the much wider context of the analysis of arbitrary matrices of connections (see, for example, Section 2.2).

§2. The approximation of graphs

1. Approximation problems in a space of relations. In the preceding section we established that to analyze interallelic complementation it is necessary to understand the structure of the initial complementation matrix. At first glance there is nothing at all new in this assertion; the entire preceding chapter was dedicated to the analysis of such matrices. However, then we had an exact idea of the type of structures to be expected in the real system: a linear or near-linear ordering of its elements. Since the mosaic model does not provide any information about the structure, a more general mathematical language is required.

Before giving an exact statement of the problem we will briefly discuss the general notion of structure in a system of interconnected objects.

The analysis of a system's structure is usually understood to mean the representation of its objects and their basic connections in an aggregate form. For example, structures of an ordered type are possible (such as were studied in the preceding chapter), when connections are defined according to the nearness of objects in some ordering. Also possible are structures associated with partitions, in which the system is defined in the form of a collection of relatively unconnected groups of objects, with the basic connections of the system being concentrated within the groups. Structures of more complex forms are possible, involving cycles, trees etc.

This brings us to the following more precise notion. We say that a *partition* $R = \{R_1,\ldots,R_m\}$ on a set A is *structured* with structure $\varkappa$, if $\varkappa$ is an oriented graph (relation) on the set $\{1,\ldots,m\}$ of indices of the classes of the partition R. The fact that $(s,t) \in \varkappa$ is interpreted as the existence of an essential (though not necessarily bidirectional!) connection between the classes R_s and R_t.

The pair $(R,\varkappa)$ is sometimes called the *macrostructure* of the set A, since it is an aggregated description of the system.

If $\varkappa = \phi$, then $(R,\varkappa)$ is nothing more than the usual unordered partition of the set A. If $\varkappa$ is a linear order then $(R,\varkappa)$ is an ordered partition. If $\varkappa$ is a cycle then $(R,\varkappa)$ is a cyclic partition, etc.

To each partition with structure $(R,\varkappa)$ there corresponds a relation $R \subseteq A\times A$, defined by the condition

$$(i,j) \in R_{\varkappa} \longleftrightarrow \quad \exists s,t=1,\ldots,m$$

$$[i \in R_s \text{ and } j \in R_t \text{ and } (s,t) \in \varkappa].$$

It is clear that the relation R on A coincides with $R_{\varkappa}$ if and only if there exists a homomorphism of the relation $R \subseteq A\times A$ onto the relation $\varkappa \subseteq \{1,\ldots,m\}^2$. This homomorphism is specified by associating with each object $i \in A$ the number of the class of R which contains it.

We consider the inverse problem of reconstructing the macrostructure corresponding to a given relation $R \subseteq A\times A$. Clearly, its solution is the partition into groups of objects R_s with identical images and pre-images, specified by the canonical homomorphism. The corresponding structure is defined by the condition: $(s,t) \in \varkappa — i \in R_s, j \in R_t \; [(i,j) \in R]$, which is correct, since if there exist in R_s and R_t objects i,j such that $(i,j) \in R$, then $R_s \times R_t \subseteq R$, i.e. every pair of objects $(i,j) \in R_s \times R_t$ is contained in R.

However this solution is not unique, since any partition more detailed than R, and with similarly defined structure is also generated by the homomorphism, and consequently forms a macrostructure corresponding to R.

The matrix corresponding to the relation $R_{\varkappa}$ (i.e. the partition with structure $(R,\varkappa)$), is defined as the $N\times N$ matrix consisting of $m\times m$

cells $r_{st} = \|r_{ij}\|_{i\varepsilon R_s, j\varepsilon R_t}$, all elements of which have the same value: 0, if $(s,t)\notin\varkappa$, and 1, if $(s,t)\varepsilon\varkappa$.

In particular, the matrix associated with an unordered partition has unit cells on the main diagonal (and only there), and the matrix associated with an ordered partition has as its only zero blocks those situated below the main diagonal. The matrix associated with a strict ordering of the objects can be reduced, by a simultaneous permutation of rows and columns, to a triangular form, in which all elements above the main diagonal are ones and all elements below it are zeros.

Thus any type of structure which can be described in terms of partitions with structure $(R,\varkappa)$ is characterized by binary relations (graphs) on A, or equivalently, by $N\times N$ Boolean matrices.

Let E be the class of relations (matrices) on A, which characterize the hypothetical types of structures for a given system. For instance it could be the class of rankings (linear or interval orderings), as was the case in the preceding chapter, when we spoke of a known, linear type of structure, with only the mutual disposition of elements being unknown. It could also be the class E_m of matrices corresponding to all possible macrostructures whose number of classes does not exceed m, if, as was the case with interallelic complementation of mutations for a specific gene, there is no basis for preferring one type of structure over another, but only the confidence that the number of functional centers does not exceed m.

Our task is to find in E that relation (matrix) which approximates the given complementation matrix (or the relation corresponding to it) closely enough. In the most simple situation E contains a macrostructure $(R,\varkappa)$ with a relation $\varkappa$ which is homomorphic to the original complementation relation I: it characterizes the macrostructure of the matrix of the complementation data, since the corresponding matrix $R_\varkappa$ coincides with the original one. It would seem to be exactly this situation for interallelic complementation, if the postulated mosaic mechanism is correct. For if the centers interact, then any mutation affecting one of them automatically alters the others. Therefore if $(i,j)\varepsilon I$ for i and j from different centers, this will also be true for any other i, j from these centers.

However this is not so. Even though a mutation affects a center as a whole, whether the effect is transmitted to another center which interacts with the first is not clear. It all depends on the extent to which the defect influences the interaction of the centers. Therefore a real complementation matrix is not homomorphic to the structure of the interactions among its functional centers. One can at least say that it must be nearly so.

Therefore to select a "representative" of the initial matrix in the class E it is necessary to learn to measure the degree of nearness between matrices. A natural measure of closeness of two matrices $r = \|r_{ij}\|$ and $p = \|p_{ij}\|$ is the *Hamming distance*, the number of corresponding positions in which they differ:

$$d(r,p) = \sum_{i,j=1}^{N} |r_{ij}-p_{ij}| = \sum_{i,j=1}^{N} (r_{ij}-p_{ij})^2. \tag{1}$$

The last equality in formula (1) follows from the important property of Boolean quantities, that they coincide with their squares: $0^2 = 0$ and $1^2 = 1$.

In terms of relations the distance (1) is calculated as the number of pairs (i,j) which are contained in one and only one of them. In terms of diagrams the distance is the minimal number of operations of adding and deleting arcs necessary to change one graph into the other. Several properties of this distance measure are studied in the books [21,22], where it is used in the analysis of qualitative phenomena. Its importance for us lies in its possible use for selecting from E that relation which is closest to the initial matrix.

In the next section we will also consider the case when a relation is sought in E which is nearest to several given relations.† We therefore examine the general formal problem.

Let $R^1, \ldots, R^n$ be n given relations on the set A. It is required to approximate this system of relations by a relation from the class E, i.e. to find a relation $R \varepsilon E$ for which the sum

$$f(R) = \sum_{k=1}^{n} d(R,R^k) \tag{2}$$

is minimal, for all $R \varepsilon E$.

This approximating relation is naturally called the *median* of the system $\{R^1, \ldots, R^n\}$ in the class E.

The following assertion is fundamental to the material we will be discussing [20].

<u>Theorem 1</u>. A relation R is the median of the system $\{R^1, \ldots, R^n\}$ in the class E if and only if it maximizes

$$g(R) = \sum_{(i,j)\varepsilon R} b_{ij} = \sum_{s,j=1}^{N} b_{ij} r_{ij} \qquad (R \varepsilon E),$$

where

$$b_{ij} = a_{ij} - \frac{n}{2}, \qquad a_{ij} = \sum_{k=1}^{n} r_{ij}^k.$$

---

†These situations also arise frequently in such data analysis tasks as the mutual reduction of expert orderings or the construction of classifications based on qualitative traits.

Proof. We consider the function $f(R)$ which, from (1), is equal to

$$f(R) = \sum_{i,j=1}^{N} \sum_{k=1}^{n} (r_{ij}+r_{ij}^{k}-2r_{ij}r_{ij}^{k}).$$

Noticing that $\sum_{k=1}^{n} r_{ij} = nr_{ij}$, $\sum_{k=1}^{n} r_{ij}^{k} = a_{ij}$, we have

$$f(R) = \sum_{i,j=1}^{N} a_{ij} - 2\sum_{i,j=1}^{N} \left(a_{ij}-\frac{n}{2}\right) r_{ij}.$$

Since $\sum_{i,j=1}^{N} a_{ij}$ is fixed for given $R^1, \ldots, R^n$ the minimization of $f(R)$ is equivalent to the maximization of the second member of the sum. Q.E.D.

2. The optimal partition problem [15]. We consider the approximation problem in the specific case when E is the set of all (unordered) partitions $R = \{R_1, \ldots, R_m\}$ of the set A (for arbitrary $m \geq 1$). For a matrix r corresponding to the partition $R = \{R_1, \ldots, R_m\}$, an element r_{ij} equals 1 if and only if for some s ($s=1,\ldots,m$) the objects i and j both belong to R_s. Consequently, in this case

$$g(R) = \sum_{s=1}^{m} \sum_{i,j \in R_s} \left(a_{ij}-\frac{n}{2}\right). \tag{3}$$

Expanding the bracketed part of (3) we have

$$g(R) = \sum_{s=1}^{m} \sum_{i,j \in R_s} a_{ij} - \frac{n}{2}\sum_{s=1}^{m} N_s^2, \tag{4}$$

where N_s is the number of objects in the class R_s.

Equation (3) has a simple interpretation. The quantities $a_{ij}=\sum_{k=1}^{n} r_{ij}^{k}$ are naturally interpreted as indicators of connection of the objects i and j in the system $R^1, \ldots, R^n$. Therefore $g(R)$ in (3) is simply the sum of the "internal" connections (under subtraction from each of them of the significance "threshhold" $\frac{n}{2}$) in the partition R. Solving the approximation problem then reduces to constructing the partition for which the sum (3) of internal connections with subtracted significance threshhold is maximized. The meaning of the threshhold $\frac{n}{2}$ becomes clear in formula (4): To maximize $g(R)$ in agreement with (4) it is necessary to increase the sum of internal connections $\sum\sum a_{ij}$, and to decrease the sum of squares of class sizes $\sum N$. Maximizing the first member of (4) increases the "compactness" of the partition R, while minimizing the second member increases the degree of uniformity of its distribution.

Actually, the minimum $\sum N_s^2$ is reached for $N_1 = \dots = N_m$, since $\sum_s N_s = N$ is fixed. Thus the size of the threshhold $\frac{n}{2}$ characterizes the degree of compromise between the two non-coincident goals of "compactness" and "uniformity" of the desired classification.

For $n=1$, when a single initial relation R^1 with matrix $\|r_{ij}^1\|$ is being approximated, the elements of the matrix $\|a_{ij}-\frac{n}{2}\|$ are the oppositely placed numbers $1/2$ and $-1/2$. Then the elements of the matrix $\|2a_{ij}-\frac{n}{2}\|$ are 1 and -1. The approximation problem in a class of equivalence relations is solved by maximizing $2g(R)$, the sum of the internal "connections" of the form $2a_{ij}-1$ in the partition R. A symmetric matrix with elements 1 and -1 is frequently associated with the so-called *signed graph* [10], a complete non-oriented graph, each edge of which is marked with the sign "+" or "-". Thus in order to approximate a given graph it necessary to pass over to the corresponding signed graph, in which an edge is "positive" if it appears in the initial graph, and "negative" otherwise, and then introduce an optimal cut of the signed graph, so that the edges are "cut" in such a way as to minimize the algebraic sum of the losses.

The exact solution of this problem is extremely laborious, even for a small number of objects. We introduce without proof some results of G. S. Friedman[44] which help to limit the number of trials, in some cases.

If a regular graph does not contain a triangle, then any maximal pair-combination of the graph [7] solves the approximation problem.

If the maximal length of a non-repeating chain (cycle) of the initial graph does not exceed $k<N$, then the cardinalities of the classes of an optimal partition do not exceed k (respectively $k+1$).

The number of classes of any optimal partition is not less than two-thirds of the radius of the initial graph. If the initial graph has a k-element clique then any optimal partition has a class with no fewer than $\left[\frac{k}{2}\right] + 1$ elements.

In connection with the complementation matrix $\|a_{ij}\|$, keeping in mind that for the mosaic mechanism a single class of a partition contains those mutations which are "more or less" similar in their complementation reactions, one should consider as the indicators of a given collection of muta- their complementarities with all of the mutations under consideration. More formally, the indicators of a given collection (of mutations) are the columns of the complementation matrix: For any mutation i the jth column tells whether mutation i is complementary to mutation j, according to whether a_{ij} is zero or not. Thus functional similarity of mutations is characterized by N binary indicators corresponding to the columns of the complementation matrix. For a particular mutation each such indicator partitions the set A into two classes of mutations (in the one class are those mutations complementary to it, and in the other, those which are not).

By Theorem 1 the approximation of these N indicators can be replaced by the classification task for the matrix of connections $a_{ij} = \sum_{k=1}^{N} r_{ij}^{k}$. What does the quantity a_{ij} mean in this case? The value of r_{ij}^{k} is 1 if the mutations i and j are identically complementary with the kth mutation, i.e. if the numbers in positions i and j of the kth column of the complementation matrix are identical. Consequently a_{ij} is none other than the number of identical positions in rows i and j of the complementation matrix, viewed as N-element vectors. We will call these quantities a_{ij} *indicators of functional similarity* of mutations i and j.

Curiously, it is usual in experimental investigations involving objects represented by binary indicators, to measure similarity by the number of identical positions (or non-identical ones). We see from the above that this measure also shows up in the framework of an approximation problem.

Experiments have shown that in analyzing specific data it is useful to vary the value of the significance threshhold of individual connections (it is the indicator of the degree of compromise between the requirements of "compactness" and uniformity).

In the book [22] it is shown how to interpret the value of the threshhold when it is not necessarily $\frac{n}{2}$, in geometric terms (see also [25]). We consider the modified problem of constructing an optimal partition $R = \{R_1, \ldots, R_m\}$ according to the criterion

$$g(a,R) = \sum_{s=1}^{m} \sum_{i,j\varepsilon R_s} (a_{ij}-a) = \sum_{s=1}^{m} \sum_{i,j\varepsilon R_s} a_{ij} - a\sum_{s=1}^{m} N_s^2, \qquad (5)$$

where a is an arbitrary real number and R is one of a fixed class E of partitions on the set A [15,22].

We note that in this formulation the problem has meaning not only in the framework of approximations, but can also be applied to connection matrices $\|a_{ij}\|$ of arbitrary form.

We first convince ourselves that a partition, optimal in the sense of (5), is also "good" in a meaningful sense of that word unconnected with any indicator of optimality.

In a "good" partition each class is "good", i.e. "compact". To sharpen this aphorism we introduce the following definitions.

A partition $R = \{R_1, \ldots, R_m\}$ is *compact* if the average internal connection between objects in each class exceeds the average external connection between two classes:

$$\frac{1}{N_s(N_s-1)}\sum_{i,j\in R_s} a_{ij} \geqslant \frac{1}{N_q N_t}\sum_{i\in R_q}\sum_{j\in R_t} a_{ij} \qquad (6)$$
$$(q,s,t=1,\ldots,m),$$

where N_s is the number of objects in class R_s $(s=1, \ldots, m)$.

A set $S\subseteq A$ is a *concentrate* if the average connection between objects of S exceeds the average connection of S with the remaining objects:

$$\frac{1}{|S|(|S|-1)}\sum_{i,j\in S} a_{ij} \geqslant \frac{1}{|S|(N-|S|)}\sum_{i\in S}\sum_{j\notin S} a_{ij}. \qquad (7)$$

Although the notions introduced also have no connection with any indicator of the quality of the partition, it will be shown in what follows that a partition, optimal in the sense of (5), must be compact and that its classes must be concentrates.

We note that for the case when R_s and S contain only single elements the left parts of the inequalities (6) and (7) are meaningless. From the preceding footnote they should be taken to be 1.

As a preliminary, note that in analyzing an optimal partition in the sense of (5) for an arbitrary connection matrix, we can limit our attention to the case when the connections are symmetric $(a_{ij}=a_{ji})$. For if $\|a_{ij}\|$ is not symmetric the passage to a new symmetric matrix $\|a'_{ij}\|$ with $a'_{ij} = \frac{a_{ij}+a_{ji}}{2}$ is straight-forward, and does not change the values in (5) or the optimal partition.

Now for a given partition $R = \{R_1, \ldots, R_m\}$ and number $a\geqslant 0$ let

$$A_{st} = \sum_{i\in R_s}\sum_{j\in R_t}(a_{ij}-a).$$

The quantity A_{st} is the sum of elements $a_{ij}-a$ of the connection matrix "cell" $a_{ij}-a(i\in R_s, j\in R_t)$ corresponding to classes R_s and R_t of the

---

Here and in the remainder of this chapter a connection a_{ii} of an object with itself is not considered. This means that our discussion is about approximating data in the class of Boolean matrices with all zeros on the main diagonal (graphs without loops). Such restrictions will not be stated formally. In particular, removing the parentheses in (3) we obtain in (4) the expression $\sum N_s(N_s-1)$ in place of $\sum N_s^2$, since these quantities differ by a constant (equal to N), we may, as before, consider (4), and more generally (5), as optimization indicators. Similarly, since for $r_{ii}=0$ $a_{ii}r_{ii}=0$, in the sums $\sum a_{ij}$ one need not write the restriction $i\neq j$, but consider that $a_{ii}=0$. This corresponds to the fact that in the initial matrices r^k one considers elements on the main diagonal to be zeros.

partition R. Since we are not considering the connection of an object with itself it is natural to take $A_{ss}=0$ for a single-element class R_s.

The following necessary condition for the optimality of a partition holds [15]:

Theorem 2. If a partition $R = \{R_1, \ldots, R_m\}$ is optimal according to (5) in the class of all possible partitions of the set A, then for $s \neq t$ $(s,t=1,\ldots,m)$ $A_{st} \leqslant 0$, while $A_{ss} \geqslant 0$ for all s.

Proof. Suppose that R is optimal, and for some s $A_{ss}<0$ In this case we move from R to a better partition R' by "separating" the objects of class R_s into single-element classes $\{i\}$ for $i \in R_s$. But then $\sum_{i \in R_s} A_{ii}=0>A_{ss}$, so that $g(a,R')$ is obtained from $g(a,R) = \sum_{t=1}^{m} A_{tt}$ by the exclusion of $A_{ss}<0$ from the number of terms, and thus $g(a,R')>g(a,R)$. But this is contrary to the optimality of R.

On the other hand, if $A_{st}>0$ for $s \neq t$, then an increase in the value of (5) can be obtained by combining classes R_s and R_t. In this case the quantity $A_{st}+A_{ts}=2A_{st}>0$ is added to the sum, which again contradicts the optimality of R. This proves the theorem.

The required result now follows easily [15]:

Theorem 3. A partition, optimal in the sense of (5), is compact, and each of its classes is a concentrate.

Proof. From Theorem 2 the following inequalities hold for an optimal R:

$$A_{ss} = \sum_{i,j \in R_s} a_{ij} - aN_s(N_s-1) \geqslant 0,$$

$$A_{qt} = \sum_{i \in R_q} \sum_{j \in R_t} a_{ij} - aN_q N_t \leqslant 0.$$

From this it follows that

$$A_q = \sum_{t \neq q} A_{qt} = \sum_{i \in R_q} \sum_{j \notin R_q} (a_{ij}-a) = \sum_{i \in R_q} \sum_{j \notin R_q} a_{ij} - aN_q(N-N_q) \leqslant 0.$$

But then

$$\frac{1}{N_s(N_s-1)} \sum_{i,j \in R_s} a_{ij} \geqslant a,$$

$$\frac{1}{N_q N_t} \sum_{i \in R_q} \sum_{j \in R_t} a_{ij} \leqslant a,$$

$$\frac{1}{N_q(N-N_q)} \sum_{i \in R_q} \sum_{j \notin R_q} a_{ij} \leqslant a.$$

Finally, inequalities (6) and (7) follow from this, completing the proof.

We now consider the question of the relationship between an optimal partition and the value of the threshhold a [12].

We first remark that if $a_{ij} \geq a$ for all i,j i.e. $a_{ij}-a \geq 0$ for all i,j, then the largest value of (5) is reached for the partition consisting of a single class containing all the objects. On the other hand, if for all i,j $a_{ij} \leq a$, i.e. $a_{ij}-a \leq 0$, then the optimal (zero) value of the quantity $g(a,R)$ is obtained for the trivial partition which divides the set A into N single-element classes.

Changing a from the first situation ($a_{ij} \geq a$) to the second ($a_{ij} \leq a$) corresponds to changing the number of classes of the optimal partition from 1 to N. The value of the threshhold a, as well as the number of classes m, reflects the degree of "coarseness" of aggregation of the initial information.

Can we assert that the number of classes of an optimal partition is monotonically dependent on a? The answer is "no". It is not difficult to find examples in which increasing the threshhold decreases the number of classes of the optimal partition. Moreover, there exist connection matrices for which m does not take on all the values from 0 to N, for all possible values of a [12].

However, there is a second, more refined characteristic of the optimal partition, which *is* dependent on a. It is the quantity $\Delta(R) = \sum_{s=1}^{m} N_s^2$ which, as was remarked earlier, characterizes the degree of non-uniformity of the distribution of objects to the classes of R. This quantity is closely connected with the well-known measure of the non-uniformity of a distribution, entropy. It is obtained by taking the quadratic part of the Taylor expansion of the entropy (see also [22]).

<u>Theorem 4</u> [12]. For a partition R, optimal according to $g(a,R)$, the value $\Delta(R)$ is monotonically non-increasing for increasing a.

<u>Proof</u>. Let R' be optimal for $a = a'$ and R'' be optimal for $a = a''$, with $a' > a''$. We must show that $\Delta(R') \leq \Delta(R'')$.

From (5) we have

$$g(a,R) = g(0,R) - a\Delta(R).$$

By the definition of R' and R'',

$$g(a',R') \geq g(a',R''),$$

$$g(a'',R'') \geq g(a'',R').$$

Substituting the expression $g(a,R)$ in these inequalities we obtain

$$g(0,R') - a'\Delta(R') \geq g(0,R'') - a'\Delta(R''),$$

$$g(0,R'')-a''\Delta(R'') \geq g(0,R')-a''\Delta(R').$$

Hence

$$a''(\Delta(R')-\Delta(R'')) \geq g(0,R')-g(0,R'') \geq a'(\Delta(R')-\Delta(R'')).$$

Consequently,

$$a''(\Delta(R')-\Delta(R'')) \geq a'(\Delta(R')-\Delta(R'')),$$

i.e.

$$(a - a'')(\Delta(R')-\Delta(R'')) \leqslant 0.$$

Since $a'-a''>0$, this is possible only for

$$\Delta(R')-\Delta(R'') \leqslant 0,$$

Q.E.D.

Notice that for the proof only the optimality of R' and R'' was used, independently of the class E of admissible relations. This means that Theorem 4 is in fact true for approximative relations from any class of relations E by utilization of the quantity $\sum_{(i,j)\in P}(a_{ij}-a)$ $(P\in E)$: with increase of a the value $\Delta(P)=|P|$ decreases. This is true in particular in the case when E is a collection of partitions with a previously fixed number of classes.

Searching for an optimal partition (in the class of all possible partitions), which is minimal in the sense of (5), is a combinatorily complex task, and in general, it cannot be solved by any simple methods. In these situations a frequent approach is to build a simple algorithm which gives an exact solution in the large majority of cases, but not necessarily in all cases. Unfortunately, in seeking an optimal partition this approach is not suitable, as the following result shows.

We will say that a property of systems of relations $\{R^1, \ldots, R^n\}$ holds *almost always* if the fraction of collections $\{R^1, \ldots, R^n\}$ which satisfy the property approaches 1 as N approaches ∞.

<u>Theorem 5</u>. The summed distances $f(R)$ from a system of relations $\{R^1, \ldots, R^n\}$ to an optimal equivalence R is asymptotically equal to $\frac{nN^2}{2}$ almost always, i.e.

$$\frac{f(R)}{\frac{N^2n}{2}} \longrightarrow 1 \quad (N\longrightarrow\infty).$$

The proof of the theorem follows from a result of G. S. Friedman, who showed that the distance from an arbitrary relation to the equivalence nearest it is almost always asymptotically equal to $\frac{N^2}{2}$. This means that for given N the collection of relations for which the distance from the nearest equivalence is not less than $(1-\delta_N)\frac{N^2}{2}$ contains a fraction not less than $(1-\varepsilon_N)$ of the whole quantity of relations, with $\delta_N, \varepsilon_N \longrightarrow 0$ for $N \longrightarrow \infty$ [44]. Designate the size of this class by $\overline{N}$, and consider all possible n-tuples of relations from this class. The number of these is $C^n_{\overline{N}+n-1}$ from the well-known formula for the number of combinations of things with repetitions. From the above,

$$C^n_{\overline{N}+n-1} > C^n_{(1-\varepsilon_N)2^{N^2}+n-1},$$

since the entire number of relations is 2^{N^2}. But then the fraction of the n-tuples being considered exceeds

$$\frac{C^n_{(1-\varepsilon_N)2^{N^2}+n-1}}{C^n_{2^{N^2}+n-1}}$$

and, clearly approaches 1 as $N \longrightarrow \infty$ for fixed (or even slowly increasing) n.

For each relation from these n-tuples the distance to the nearest equivalence is asymptotically equal to $\frac{N^2}{2}$, so that the sum of the distances is asymptotically equal to $\frac{nN^2}{2}$. Q.E.D.

We note in addition that there are $\frac{N^2}{2}$ arcs in the median relation, and for $N \longrightarrow \infty$ almost all the relations have asymptotically the same number of arcs. Consequently, from Theorem 5 it follows that for almost all collections $\{R^1, \ldots, R^n\}$ the median in the class of partitions may be obtained by rejection of all arcs in $\{R^1, \ldots, R^n\}$, i.e. by taking the trivial partition, or vice versa, adding arcs necessary to make the universal relation. This means that any relation is almost always asymptotically optimal, including the trivial relation consisting of N classes, and the universal relation consisting of only one class. Thus, in the problem being considered, it is necessary to build locally optimal algorithms having purely experimental bases.

These algorithms work as follows. At step k some partition R of objects is considered, along with some collection of partitions $O(R)$, which constitute a "neighborhood" of R. An optimal partition is selected from this neighborhood according to the given criterion, and this becomes the new R for the next step. The algorithm terminates when the new partition coincides with the preceding one.

We will now describe two of these algorithms [22,15].

The *"unification" algorithm*. Let $R(l) = \{R_1, \ldots, R_l\}$ be an arbitrary partition of the set A into l classes. Take as the neighborhood

$O(R(l))$ all possible partitions R^{st} with $l-1$ classes, obtained from $R(l)$ by unifying the classes R_s and R_t $(s,t=1,\ldots,l)$. The optimal partition from among these becomes $R(l-1)$.

The algorithm, beginning with the trivial partition of A into N single-element classes, produces a sequence of partitions $R(N-1)$, $R(N-2)$, ...

To make use of the indicator (5), calculations are organized on the basis of the fact that $g(a,R^{st})$ is large when the sum of connections $A_{st} = \sum_{j \in R_s} \sum_{j \in R_t} (a_{ij}-a)$ is large, as follows.

For the partition $R(l) = \{R_1, \ldots, R_l\}$ consider the matrix of summed connections $\|A_{st}\|$ between its classes (at the first step, for $R(N)$ the initial matrix $\|A_{ij}=a_{ij}-a\|$ is taken). The maximal element A_{st} of this matrix is selected. If it is positive then the classes R_s and R_t are combined, and the transition to the $(l-1)\times(l-1)$ matrix is made, by the component-by-component composition of the rows (and columns) s and t. From Theorem 3, the calculations are stopped when all of the A_{st} $(s\neq t)$ are non-positive.

The partition thus obtained satisfies the necessary optimality condition, and consequently it is compact and its classes are concentrates.

The unification algorithm, in connection with the criterion (5) is one of a widely-known group of so-called *agglomerative algorithms* of a taxonomy, consisting of successive unifications of classes near to one another. In the given case nearness of classes is estimated quantitatively. In other cases nearness of classes is estimated by: a) the value of the maximal connection between their elements (the nearest-neighbor method), b) the value of the minimal connection between their elements (the farthest-neighbor method), c) the value of the average, etc. In addition to the final partition, the whole sequence of partitions is often used in the analysis, and is viewed as a hierarchical tree (see Section 3.1.3).

The "*transfer*" *algorithm*. For any partition R, the neighborhood considered is the collection of all partitions R^{is} $(i=1,\ldots,N; s=1,\ldots,m)$, obtained by transferring the object i into the class R_s. In particular, if $i \in R_s$ then $R^{is} = R$.

At each step i is fixed, so that the optimal variant of the transfer of i into some class of R can be selected. To make use of the indicator (5) for this, it is sufficient to calculate the summed connections of i with each of the existing classes. Objects are selected in the order of their indices, with the Nth one being followed by the first. The number of these cycles can be fixed in advance.

As a rule we use a combination of the "unification" algorithm and the "transfer" algorithm in practical computations, as follows [15].

By means of the unification algorithm obtain a partition $R(l)$ with a previously fixed (and sufficiently large) number of classes. This partition is then improved by means of the transfer algorithm. The necessary condition $A_{ss} \geqslant 0$ for optimality may be violated in this process. Any such classes R_s are divided into single-element subclasses, for which the internal connections A_{ss} are zero. After this, unification produces a new partition into $l-1$ classes, which is then improved by the transfer algorithm with subsequent testing of the condition $A_{ss} \geqslant 0$ and the "dividing" of "bad" classes. Thus a sequence of alternate unifications and transfers leads to a final partition with $A_{st} \leqslant 0$ $(s \neq t)$, $A_{ss} \geqslant 0$.

3. Detecting macrostructure. From Theorem 1 the problem of detecting the macrostructure of a given $N \times N$ connection matrix $\|a_{ij}\|$ (in the simplest case $\|a_{ij}\|$ is an $N \times N$ Boolean matrix) may be formulated as follows.

Find that macrostructure $(R, \varkappa)$ in a given class of macrostructures, which maximizes the following quality indicator (for given a):

$$F(R,\varkappa) = \sum_{(s,t)\in\varkappa} \sum_{i\in R_s} \sum_{j\in R_t} (a_{ij}-a). \tag{8}$$

If $\|a_{ij}\|$ is a Boolean matrix, then by Theorem 1 $a=\frac{1}{2}$ (in the approximation problem).

For a fixed partition $R = \{R_1, \ldots, R_m\}$ a structure $\varkappa$, optimal in the sense of (8), is determined as follows. As in the preceding section, we define the quantity A_{st} to be the total connection of the classes R_s and R_t $(s,t=1,\ldots,m)$, and the structure $\varkappa(R)$ is defined by the condition:

$$(s,t)\in\varkappa(R) \longleftrightarrow A_{st}>0.$$

Clearly $F(R,\varkappa(R))$ is maximal (for all $\varkappa$). Consequently, if a pair $(R^*,\varkappa^*)$ maximizes (8), then $\varkappa^* = \varkappa(R^*)$ (to within pairs (s,t) for which $A_{st}=0$). This means that criterion (8), if all structures $\varkappa$ are admissible, forces the maximization of the sum of the positive connections of A_{st} $(s,t=1,\ldots,m)$. Since the sum of all the numbers $a_{ij}-a$ $(i,j=1,\ldots,m)$ is fixed, then during this process the maximum of the sum of the absolute values of the negative connections of A_{st} is reached.

We have thus proved that the maximization of the indicator F in the case when for every R all structures are admissible, is equivalent to the construction of a partition R for which the maximum sum of the modulus of the connections A_{st} is reached [14]:

$$F'(R) = \sum_{s,t=1}^{m} \left| \sum_{i\in R_s} \sum_{j\in R_t} (a_{ij}-a) \right|. \tag{9}$$

In other words, in optimizing the indicator (8) a partition R is obtained for which the matrix of "cells" $\|a_{ij}\|$ $(i\varepsilon R_s, j\varepsilon R_t)$ has the most contrast: The positive connections $a_{ij}-a$ are located basically in certain cells, and the negative ones in other cells. Thus objects which are more or less identical in their interactions with other objects fall into one class. Therefore the detection of macrostructure may be considered as the aggregation of information in such a way that objects which possess identical interactions are combined [14].

We designate by F^*_m the optimal value of the indicator (8) in the set of all macrostructures $(R,\varkappa)$ with a fixed number m of classes in the partition R.

Clearly,

$$F^*_m \leqslant F^*_{m+1} \quad (m=1,\ldots,N-1). \tag{10}$$

For if $F(R,\varkappa) = F_m$ and $R = \{R_1, \ldots, R_m\}$, then it is not difficult to construct a partition with $m+1$ classes, by dividing any class R into two subclasses. For this new partition the structure $\varkappa$ is redefined only for the new classes, in such a way that these classes are biconnected with each other, and also with those classes of R with which the initial class was connected. The macrostructure thus obtained clearly corresponds to a value of F equal to F^*_m, which proves inequality (10).

From (10) it follows that in the class of all possible macrostructures the optimal macrostructure is defined simply by the trivial partition R into single-element classes, with

$$(\{i\},\{j\})\,\varepsilon\,\varkappa \longleftrightarrow a_{ij} \geq a.$$

In this case $\varkappa$ is known as the *a-similarity graph* for the connection matrix $\|a_{ij}\|$.

The inequality (10) shows that, unlike the optimal partition problem, besides the threshhold a it is necessary to fix in advance the number of classes, for otherwise the a-similarity graph will be obtained. As a rule, this graph is not of interest, since it starts out with N orders [15,20].

To maximize (8) one can use the *"Structure" algorithm* [14], which is a combination of the unification and transfer algorithms described in Section 4.

In the unification algorithm, as in the maximization of (5), calculations are organized according to the matrix $\|A_{st}\|$ of the total connections between classes. However, in this case the unification of the classes R_s and R_t is realized not according to the maximum A_{st}, but by the minimum of the function

$$\Phi(s,t) = \sum_{\substack{k=1\\k\neq s,t}}^{l}\left[\left|\frac{\mathrm{sign}A_{sk}-\mathrm{sign}A_{tk}}{2}\right|\min(|A_{sk}|,|A_{tk}|)+\right.$$

$$\left.+\left|\frac{\mathrm{sign}A_{ks}-\mathrm{sign}A_{kt}}{2}\right|\min(|A_{ks}|,|A_{kt}|)\right]+\min(A_+,|A_-|), \qquad (11)$$

where A_+ (A_-) is the sum of all positive (negative) numbers from the quadruple A_{ss}, A_{st}, A_{ts}, A_{tt}.

The reasonableness of using this particular $\Phi(s,t)$ is ensured by the following equality [14]:

$$F'(R) - F'(R^{st}) = 2\Phi(s,t). \qquad (12)$$

From (12) it follows that a minimal decrease in the value of the indicator (9) as a result of unification, corresponds to the minimum of $\phi(s,t)$, so that the optimal R^{st} is actually defined by this minimum.

We now prove (12). First note that for two quantities x and y having opposite signs ($xy<0$), the following equality holds:

$$|x+y|=|x|+|y|-2\min(|x|,|y|). \qquad (13)$$

In fact, if, for example, $|x|<|y|$, then

$$|x+y|=|y|-|x|=|x + y|-2\min(|x|,|y|).$$

We continue the proof of (12) by considering the absolute values of the total connections A^{st}_{pq} between classes of the partition R^{st}. For ease of working, the class $R_s \cup R_t$ of the partition R^{st} will be designated by $R_{s\cup t}$. This permits the use, for the remaining classes of R^{st}, of the indices p, q, which they had in the partition R. We consider all possible cases:

1. $p\neq s\cup t$, $q\neq s\cup t$;
2. $p=s\cup t$, $q\neq s\cup t$;
3. $p\neq s\cup t$, $q=s\cup t$;
4. $p=s\cup t$, $q=s\cup t$.

In case 1 the classes with numbers p, q in the partitions R and R^{st} coincide, so that A^{st}_{pq} is equal to A_{pq} (the total connection of the pth and qth classes in the partition R), and of course $|A^{st}_{pq}| = |A_{pq}|$. The quantity A_{pq} does not appear in $\Phi(s,t)$, so that the "contribution" of the cell (p,q) in $F'(R)$ and $F'(R^{st})$ satisfies (12).

In case 2 $p = s\cup t$, so that $A^{st}_{pq} = A_{sq} + A_{tq}$. If the signs of A_{sq} and A_{tq} agree, then $|A^{st}_{pq}| = |A_{sq} + A_{tq}| = |A_{sq}| + |A_{tq}|$, and consequently the "contribution" of the cell (p,q) in the left part of (12) is zero. This cell also does not introduce anything into $\Phi(s,t)$, since $\text{sign}A_{sq} - \text{sign}A_{at} = 0$, i.e. the coefficient for the corresponding term $\min_i(A_{sq},A_{tq})$ is zero. On the other hand, if $\text{sign}A_{sq} \neq \text{sign}A_{tq}$, then from (13) $|A_{pq}^{st}| = |A_{sq}| + |A_{tq}| - 2\min(|A_{sq}|,|A_{tq}|)$. Consequently $|A_{sq}| + |A_{tq}| - |A_{pq}^{st}| = 2\min(|A_{sq}|,|A_{tq}|) = |\text{sign}A_{sq} - \text{sign}A_{tq}|\min(|A_{sq}|,|A_{tq}|)$. So that in this case also the "contribution" of the cell (p,q) in the partition R^{st} is identical in the left and right parts of (12).

Case 3 is handled similarly.

In case 4 $p = s \cup t$, $q = s \cup t$, so that $A_{pq}^{st} = A_{ss} + A_{st} + A_{ts} + A_{tt}$. In accord with (11) we designate by A_+ the sum of the positive numbers among the quadruple A_{ss}, A_{st}, A_{ts}, A_{tt}, and by A_- the sum of those that are negative. We know that $|A_{ss}| + |A_{st}| + |A_{ts}| + |A_{tt}| = |A_+| + |A_-|$. On the other hand, $|A_{pq}^{st}| = |A_+ + A_-| = |A_+| + |A_-| - 2\min(A_+,|A_-|)$, i.e. $|A_{ss}| + |A_{st}| + |A_{ts}| + |A_{tt}| - |A_{pq}^{st}| = 2\min(A_+,|A_-|)$, so that again the contributions of the cell (p,q) to the left and right sides of (12) are equal.

This concludes the proof of the equality (12).

Thus, beginning with the trivial partition into N single-element classes, it is possible to successively combine pairs of classes by the unification algorithm in accordance with the minimum $\Phi(s,t)$.

The transfer algorithm is applied just as described on page 113, except that the valuation of the partition R^{is} obtained from R by transferring object i from R^t $(t=1,\dots,m)$ into class s, is not produced according to the sum of the connections of i with the objects of R^s, but according to the maximum value of

$$f(i,t,s) = \sum_{k\in\varkappa(s)\cup\varkappa^{-1}(s)} \sum_{j\in R_k} (a_{ij}-a) - \sum_{k\in\varkappa(t)\cup\varkappa^{-1}(t)} \sum_{j\in R_k} (a_{ij}-a). \quad (14)$$

Here $\varkappa = \varkappa(R)$ and the first summand characterizes the connection of the object i within the macrostructure after its transfer from R_t to R_s, and the second summand, the connection of i before the transfer.

As in the search for a partition, the best results are obtained by a combination of the unification algorithm and the transfer algorithm.

The value of the threshhold a can be determined from considerations of interest. In [25] the method of least squares is also used to determine the value of a for which the desired macrostructure of the "best form" approximates a given connection matrix $\|a_{ij}\|$, whether or not it is Boolean.

§3. Analyzing the spatio-functional organization of specific genetic systems

1. Complex protein organization. The goal of this section is, on the one hand, to examine the validity of the mosaic mechanism of inter-allelic complementation, and, on the other, to obtain individual descriptions of the functional organization of specific systems of "gene-enzymes."

The "structure" algorithm was used in analyzing several specific cistrons. We will examine here the results of only one of these. This cistron is studied both genetically and biochemically more throughly than the others in order to form a possible molecular interpretation of the constructed graphs.

The discussion will be about the B cistron of the his operon, which controls the synthesis of the amino acid histidine in *Salmonella* bacteria (see Fig. 17). A peculiarity of this cistron is that its protein product, the enzyme, dehydratase-phosphatase, is bifunctional: it catalyzes two non-adjacent reactions (the 7th and 9th) in the biosynthesis of histidine. The reactions are: the splitting off of water molecules (*dehydratase function*) and the adjoining of a phosphoric-acid residue (*phosphatase function*). Correspondingly, some mutations may be defective for the dehydratase function, some for the phosphatase function, and some for both functions. However, in spite of the large number of complementary mutations which have been described in this cistron, up to this time only the defects for the phosphatase activity have not been located [71,81,48]. The results below allow, in particular, an explanation of this peculiarity (see page 125).

The initial complementation data were taken from the work of Loper et. al. [81]. Out of 64 his B mutants which were studied, some coincided in their complementation behavior, i.e. in the image and preimage of the corresponding relation, so that the complementation matrix (Fig. 26), with which all further operations were carried out, consisted of 36 different groups of mutants. (Clearly, these groups are none other than the classes of the canonical partition; see the Appendix.)

In the given collection of mutations there are various types of changes: The greater part are the usual point replacements "codon⟶codon", replacements of the type "codon⟶nonsense", frameshift (i.e. the loss and insertion of individual bases), and deletions of varying extents. Some of these are defective only for the dehydratase activity (the monofunctional ones underlined in Fig. 27,b), some for both activities (the bifunctional ones written in parentheses in Fig. 27,b), and for the remaining mutations the functions affected by them are still not known.

The results of aggregating mutations by means of the "structure" algorithm are shown in Figure 27. It was natural to select as the value

	12,...	40,...	53,...	56	59	65	116	136,...	138	167,488	206,...	217	234	241	243	262	286	289	353	355	380	391,836	470	480	542	562,812	569	578	612	641	662	669,672	824	865	902	923
12,...	1														1							1					1		1						1	
40,...		1				1						1	1	1		1		1						1	1			1								
53,...			1						1		1		1		1	1							1							1						1
56				1	1			1	1	1			1		1	1	1				1		1									1				
59				1	1			1	1	1			1		1	1	1			1	1		1			1						1		1		
65,...		1				1			1	1		1	1	1	1	1	1	1	1	1	1			1	1			1	1			1		1		
116							1	1	1	1			1		1	1	1			1	1		1			1						1		1		
136,...				1	1		1	1	1	1			1		1	1	1			1	1		1			1						1		1		
138			1	1	1	1	1	1	1	1		1	1	1	1	1	1	1	1	1	1		1			1						1		1		1
167,488				1	1	1	1	1	1	1			1	1	1	1	1		1	1	1		1			1						1		1		
206,...			1								1		1		1	1													1	1						1
217		1				1			1			1	1	1	1	1		1	1		1			1	1			1	1			1				
234		1	1	1	1	1	1	1	1	1	1	1	1	1	1	1	1	1	1	1	1		1			1		1	1	1		1		1		1
241		1				1			1	1		1	1	1	1	1	1	1	1	1	1			1	1			1	1			1	1	1	1	
243	1		1	1	1	1	1	1	1	1	1	1	1	1	1	1	1		1	1	1		1			1			1	1		1		1	1	1
262		1	1	1	1	1	1	1	1	1	1	1	1	1	1	1	1	1	1	1	1		1	1		1		1	1	1		1		1		1
286				1	1	1	1	1	1	1			1	1	1	1	1			1	1		1			1						1		1		
289		1				1			1			1	1	1		1		1	1		1			1	1			1	1				1		1	
353						1			1	1		1	1	1	1	1		1	1		1							1				1		1		
355					1	1	1	1	1	1			1	1	1	1	1			1	1		1			1						1		1		
380				1	1	1	1	1	1	1		1	1	1	1	1	1	1	1	1	1		1			1						1		1		
391,836	1																					1					1		1						1	
470			1	1	1		1	1	1	1			1		1	1	1			1	1		1			1						1		1		1
480		1				1						1		1		1		1						1	1			1	1							
542		1				1						1		1				1						1	1			1	1							
562,812					1		1	1	1	1			1		1	1	1			1	1		1			1						1		1		
569	1																					1					1		1		1				1	
578		1				1						1	1	1		1		1	1					1	1			1								
612	1					1					1	1	1	1	1	1		1				1		1	1		1		1		1		1		1	1
641			1								1		1		1	1														1						1
662																											1		1		1				1	
669,672				1	1	1	1	1	1	1		1	1	1	1	1	1		1	1	1		1			1						1		1		
824														1				1											1				1		1	
865					1	1	1	1	1	1			1	1	1	1	1		1	1	1		1			1						1		1		
902	1													1	1			1				1					1		1		1		1		1	
923			1						1		1		1		1	1							1						1	1						1

Figure 26. The matrix of interallelic complementation for cistron B of the his-operon in *Salmonella typhimurium*. Mutants are numbered as in the original paper of Loper et. al. [81]. Each row (column) corresponds to a separate complementation group which is the union of mutants with identical complementation reactions. For groups containing more than one mutant, only one of them is designated (non-complementarity is designated by 1, complementarity by a blank).

of the threshhold $a=\frac{1}{2}$, so that, formally, the problem solved was that of approximating a relation given by the matrix of Fig. 26, in the class E_m of partitions with structure. Since in this case a value for a was not known in advance, the problem was solved for various a. Beginning with seven classes, it was found that each successive partition for larger a was contained in the preceding one. In other words, the transition to a larger number of classes simply produces a subdivision of classes, not their rearrangement, with the closest "neighborhood" of any mutation in the structure being unchanged. This permitted stopping at $m=7$.

	391,836	12,...	569	662	902	612	824	289	578	40,...	480	542	217	353	65,...	241	380	669,672	167,488	865	138	286	355	136,...	59	562,812	116	56	470	262	234	243	53,...	923	206,...	641
1. 391,836	1	1	1		1	1																														
2. 12,...	1	1	1		1	1																										1				
3. 569	1	1	1	1	1	1																														
4. 662			1	1	1	1																														
5. 902	1	1	1	1	1	1	1	1								1																1				
6. 612	1	1	1	1	1	1	1	1			1	1	1		1	1														1	1	1		1	1	
7. 824					1	1	1	1								1																				
8. 289					1	1	1	1	1	1	1	1	1	1	1	1	1				1									1	1					
9. 578								1	1	1	1	1	1	1	1	1														1	1					
10. 40,...								1	1	1	1	1	1		1	1														1	1					
11. 480						1		1	1	1	1	1	1		1	1														1						
12. 542						1		1	1	1	1	1	1		1	1																				
13. 217						1		1	1	1	1	1	1	1	1	1	1	1			1									1	1	1				
14. 353								1	1				1	1	1	1	1	1	1	1	1									1	1	1				
15. 65,...						1		1	1	1	1	1	1	1	1	1	1	1	1	1	1	1	1							1	1	1				
16. 241					1	1	1	1	1	1	1	1	1	1	1	1	1	1	1	1	1	1	1							1	1	1				
17. 380								1					1	1	1	1	1	1	1	1	1	1	1	1	1	1	1	1	1	1	1	1				
18. 669,672													1	1	1	1	1	1	1	1	1	1	1	1	1	1	1	1	1	1	1	1				
19. 167,488														1	1	1	1	1	1	1	1	1	1	1	1	1	1	1	1	1	1	1				
20. 865														1	1	1	1	1	1	1	1	1	1	1	1	1	1		1	1	1	1				
21. 138								1					1	1	1	1	1	1	1	1	1	1	1	1	1	1	1	1	1	1	1	1	1	1		
22. 286															1	1	1	1	1	1	1	1	1	1	1	1	1	1	1	1	1	1				
23. 355															1	1	1	1	1	1	1	1	1	1	1	1	1		1	1	1	1				
24. 126,...																	1	1	1	1	1	1	1	1	1	1	1	1	1	1	1	1				
25. 59																	1	1	1	1	1	1	1	1	1	1		1	1	1	1	1				
26. 562,812																	1	1	1	1	1	1	1	1	1	1	1		1	1	1	1				
27. 116																	1	1	1	1	1	1	1	1		1	1		1	1	1	1				
28. 56																	1	1	1		1	1		1	1			1	1	1	1	1				
29. 470																	1	1	1	1	1	1	1	1	1	1	1	1	1	1	1	1	1	1		
30. 262						1		1	1	1	1		1	1	1	1	1	1	1	1	1	1	1	1	1	1	1	1	1	1	1	1	1	1	1	1
31. 234						1		1	1	1			1	1	1	1	1	1	1	1	1	1	1	1	1	1	1	1	1	1	1	1	1	1	1	1
32. 243		1			1	1							1	1	1	1	1	1	1	1	1	1	1	1	1	1	1	1	1	1	1	1	1	1	1	1
33. 53,...																					1								1	1	1	1	1	1	1	1
34. 923						1															1								1	1	1	1	1	1	1	1
35. 206,...						1																								1	1	1	1	1	1	1
36. 641																														1	1	1	1	1	1	1

Figure 27a. Classes of complementationally close his B mutants. Reconstruction of the complementation matrix with the classes singled out.

Almost the identical partition was obtained from the construction of the optimal partition for the matrix of functional similarities of mutants with a threshhold $a=24$. This value of the threshhold corresponds to the assumption that mutations which affect the same functional centers must be identically complementary in at least *2/3* of the cases. The combination of the unification and transfer algorithms described in Section 1.2 led to a partition into seven classes, distinguished from the one shown in Fig. 27 only by the interchange of two objects and the division of class 5 into two subclasses. We will only consider further the structure of Figure 27, since the matrix of functional similarities does not precisely reflect the specific character of local paired mutational interactions, and it also forces the selection of a threshhold a for which, unfortunately, there is still no meaningful basis [14].

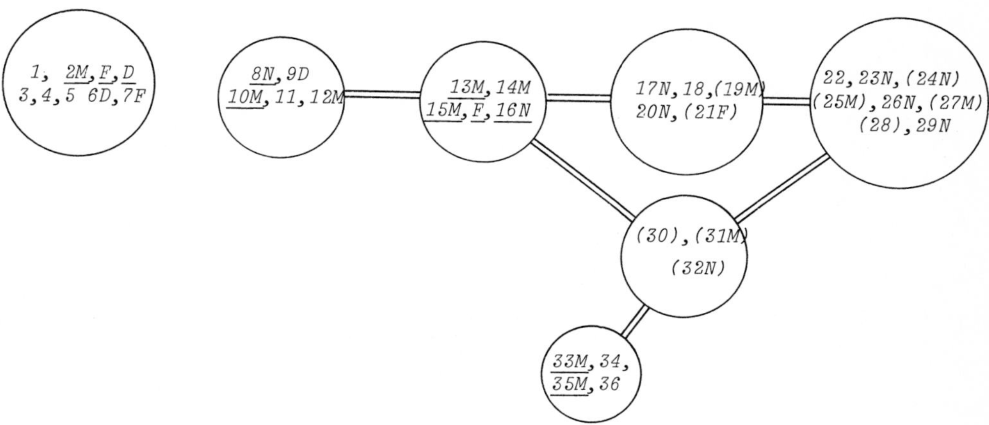

<u>Figure 27,b</u>. The classes of complementationally similar his B mutants, with inter-class connections. Most mutants are given with a two-part designation: The number is the row from Fig. 27,a and the letter indicates the type of mutation involved in each complementation group: "M" means the group contains a missense mutation, i.e. a replacement of the "codon⟶codon" type, "N"— a nonsense mutation, "F"—a frame-shift mutation and "D"—a deletion. The numbers of those groups containing monofunctional mutations are underlined, while those for groups of bifunctional mutations are enclosed in parentheses.

We will compare the structure which has been obtained with available independent data about the system.

As is evident in Fig. 27, there are no classes containing both mono-functional and bifunctional mutations in this structure. This fact is an indirect argument in support of the mosaic mechanism of interallelic complementation. Within the scope of this mechanism bifunctional mutations affect several functional centers, and consequently are identical in their complementation behavior with respect to monofunctional mutations associated with other centers. From this it is clear that classes of monofunctional mutations must be connected with one another through the class of bifunctional centers. Exactly this situation is observed in the structure of Fig. 27.

We will compare this structure with the genetic map for the his B cistron. All the classes are projected onto the cistron map (and this means onto the primary structure of the protein) non-compactly (for example, Fig. 29). This kind of disagreement between complementation and recombination has been observed in all the cistrons we have studied carefully. This is a serious argument against the model of Crick and Orgel, since in that model such discrepancies should be the exception (see also [40,43,88]). In addition, the functional centers of the protein are the results of complex spatial packing of chains and are therefore represented in the primary structure by disconnected sections.

		391,...	20M,...	569	902	662	542M	480	478D	40M,...	217M	65M,...	669,...	286	562	53M,...	923	206M,...	641
I	391,836	1	1	1	1														
	20M,...	1	1	1	1														
	569	1	1	1	1	1													
	902	1	1	1	1	1													
	662			1	1	1													
II	542M						1	1	1	1	1	1							
	480						1	1	1	1	1	1							
	578D						1	1	1	1	1	1							
	40M,...						1	1	1	1	1	1							
	217M						1	1	1	1	1	1	1						
	65M,...						1	1	1	1	1	1	1	1					
III	669,672										1	1	1	1	1				
	286											1	1	1	1				
	562												1	1	1				
IV	53M,...															1	1	1	1
	923															1	1	1	1
	206M,...															1	1	1	1
	641															1	1	1	1

Figure 28

We will now attempt to examine in a more detailed way the direct description of the functional centers of the his B protein. For this it is necessary to exclude from the initial matrix those mutations known to affect several functional centers. These are primarily the bifunctional mutations, but also a group of deletions, nonsense mutations and frameshifts. Unlike the usual point replacements, mutations of this group lead to the modification or loss of relatively long segments of the polypeptide chain. Therefore, with respect to them, it is not suitable to speak of affecting an individual functional center.

Moreover, the result of complementation testing of such mutations in pairs with missense mutations depends primarily on the ability of the corresponding polypeptide fragments to aggregate with the polypeptides carrying the missense replacements, and also consequently on the level of activity of the "mosaic" hybrid multimer. This ability frequently turns out to be allele-specific [39,88]. In particular this is pointed out by the fact that the deletions and nonsense mutations do not form a single class (see Fig. 27).

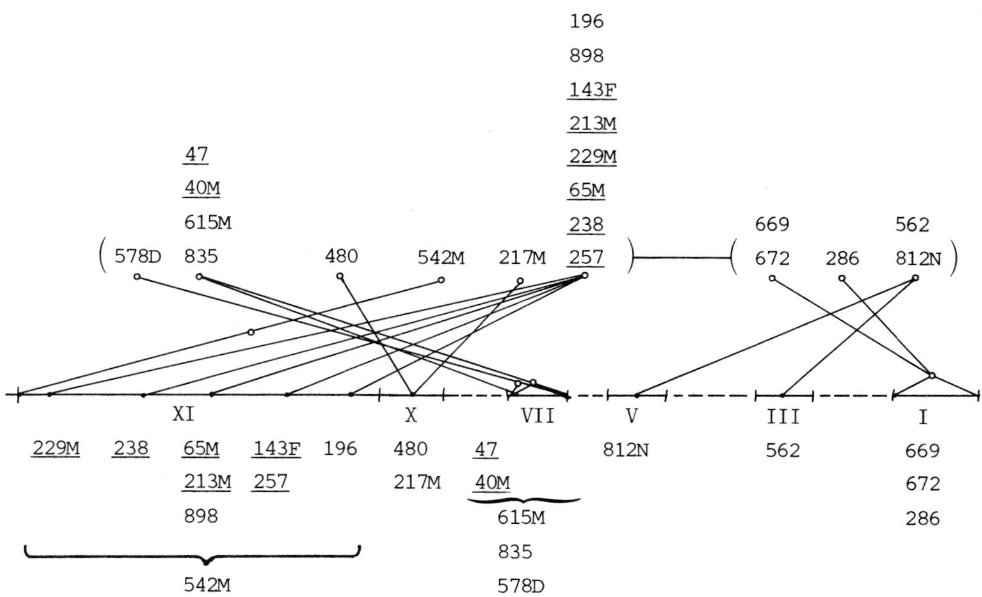

Figure 29. Projections of classes II and III (Fig. 28) on the genetic map of the cistron B: a) classes of complementationally-similar mutations, b) the genetic map of the cistron built by Hartman et. al. [71]. As is evident in this scheme, class II, which contains monofunctional mutations, is projected onto the distal (with respect to the beginning of translation in the gene) half of the map, as is also true of the other classes with monofunctional mutations. On the other hand, class III is projected onto the proximal half of the gene, where the overwhelming proportion of the bifunctional mutations are located.

As a result (Fig. 28) we obtain four almost isolated classes, three of which contain monofunctional mutations. One class (the third) may be "suspected" of bifunctionality, both from the presence in the matrix of Fig. 28 of a characteristic linkage, and from its behavior in the general structure (see Fig. 27).

The following peculiarity is associated with these classes. For monofunctional (defective for dehydratase activity) mutations, their projections onto the genetic map lie in the distant (*distal*) part of the gene, with respect to the beginning of translation. At the same time the bifunctional mutations are projected on the other, nearby (*proximal*) part of the map (see Fig. 29). Mutations of unknown nature, which are contained in a class with the known monofunctional ones, are also projected on the distal half of the map. On the other hand, the class suspected of being bifunctional is projected onto the proximal part of the map, which supports the assumption of bifunctionality of the mutations contained in it. These facts lead to the notion that the protein sections responsible for the dehydratase activity are a collection of amino-acid residues from the distal half of the cistron, while the phosphatase segments are associated with the proximal half.

Analyzing the projections of complementation classes on the genetic map, one can construct a boundary separating the sections which control the two functions even at the level of the primary structure of the protein (see Fig. 29).

These results, obtained from the complementation data [81] by means of the structure algorithm, are corroborated by data from biochemical investigations on a series of mutant forms of the his B protein, carried out by Houston [72,73]. Studying the distribution of a series of nonsense mutations in the map and comparing it with the activity of the corresponding proteins, he concluded that the his B cistron is divided into two functionally differing parts: for dehydratase and phosphatase. The boundary he found is practically the same as ours.

Such coincidence of results of two completely independent approaches inspires confidence that the entire preceding construction, connected with the detection of macrostructure in a complementation matrix, is correct, and does solve the postulated problem.

With this in mind, we can proceed to the description of the functional centers of this protein. First of all three sections are distinguishable, which are responsible for the dehydratase function. These sections (centers) correspond to classes of Fig. 28 which are projected onto the distal part of the cistron. The isolated nature of these classes shows that the corresponding centers are not directly connected with one another. However, since these centers must interact (they work on exactly the same function), this leads us to hypothesize the existence of still another functional center, an allosteric center, which is connected with all of these centers and is a "conductor" of all their interactions. Naturally, it is the presence of the catalytic phosphatase center which ensures the other protein function. Since there is no mutation defective only for the phosphatase center, this center must overlap considerably (perhaps completely) with the allosteric center. By the same token, any effect of this center must be transmitted to the dehydratase portion of the molecule. The possibility that this mechanism is realized in the synthesis process for the initial polypeptide chain cannot be ruled out. Actually, the phosphatase center is a collection of amino acid residues from the proximal part of the cistron. It is logical to suppose that in the process of translation, after obtaining a necessary length, the growing polypeptide chain is folded in such a way that it forms the phosphatase center. Any distortion of this portion of the chain may lead to such curtailment or disturbance of the subsequent growth and conformation of the protein that both its functions are lost. This quite easily explains the absence of monofunctional phosphatase mutations.

We now turn our attention to the class of Fig. 27 containing the bifunctional mutations 24,25,27 and 28. It almost fails to overlap any of the classes of monofunctional mutations. Moreover, despite the simultaneous loss of both functions in the homozygote, mutations within this class can complement one another in the heterozygote (see Fig. 27,a).

These two peculiarities cannot be simultaneously explained in terms only of dehydratase, phosphatase and allosteric centers. However, if we recall the necessity of the presence in each subunit of the multimer of a contact center, by means of which it aggregates with the other subunits, then both these peculiarities are easily interpreted. It is possible that mutations of this class affect the contact zone, which leads to the de-aggregation of the subunits in the homozygote, and consequently, to the loss of all the basic properties of the multimer. In the heterozygote specific mutual correction of such defects is possible (see Section 1.1), which also explains the possibility of complementary mutations within this class. On the other hand, the absence of direct connections of these mutations with those which are defective for the the dehydratase function shows that the contact center only partly interacts with the allosteric center.

All this may be summed up in the scheme of functional organization of this enzyme (Fig. 30).

The structure thus obtained characterizes the semantic structure of the cistron: the number of functional centers in the gene product, their nature and interconnections. In particular, both the structural and functional peculiarities of lightly studied mutations can be predicted on the basis of "which neighbors they have" in the constructed graphs. In the situation we are considering an example of this is the prediction of whether mutations are monofunctional or bifunctional, depending on which class they belong to.

In our view, the use of the mathematical techniques presented here for the analysis of structure may become a highly effective means in purely genetic studies of the semantic organization of systems of "gene-enzymes."

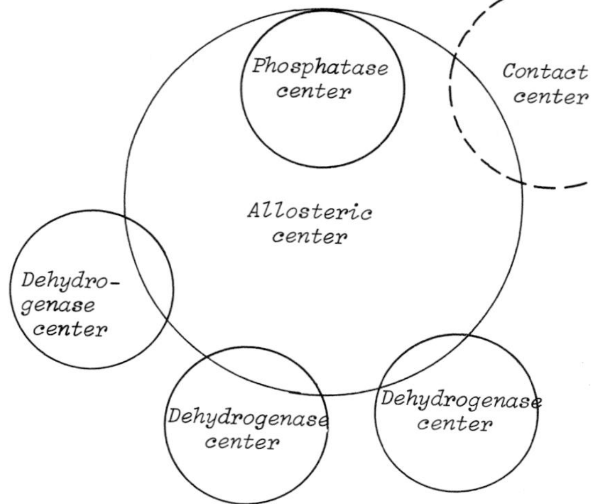

Figure 30. Scheme of functional organization of the his B cistron protein product.

2. The investigation of genome spatial structure. In the preceding investigations of specific genetic systems we restricted ourselves primarily to the analysis of genes in organisms of relatively simple structure: bacteria, viruses, bacteriophages, fungi, etc. The single exception was the locus *scute* in *Drosophila*, which, however, was naturally modeled by the most simple linear gene system of the operon type, characteristic for microorganisms.

It should be emphasized that the existence of such genetic systems in higher organisms has not been rigorously shown, up to this time. Along with such systems, one can expect essentially different methods of organization of the functional connections of gene groups.

The determination and analysis of gene systems in higher organisms by application of the usual genetic and biochemical methods which have been so fruitful for microorganisms, meets with significant difficulties. On the one hand, the complexity of the systems themselves complicates the identification of separate genetic components and their connections. On the other hand, the existing methods of experimental analysis are of little value (by reason of the much longer life cycle, as well as a number of other factors).

One might think that success in investigating complex genetic systems in higher organisms comes "from the other end", as one of the results of studying the total integrated picture of the organization of the genetic apparatus.

We examine in this regard some results concerning key elements of the spatial organization of the genetic apparatus in *Drosophila*, obtained by V. A. Kulichkov and E. F. Zhimulev, using the structure algorithm [11].

We first remark that the basic genetic processes (replication of DNA, translation and synthesis of specific proteins, mutation, recombination etc.) run their course in the cells of higher organisms not continuously, but only in a definite period of their life cycle, called the *interphase* period.

The genetic apparatus of a cell is localized primarily in the nucleus. In the interphase period the chromosomes of the nucleus are found to be in working (uncoiled) form, so that their spatial structure in the interphase nucleus differs considerably from that found during other phases. These other phases consist of the actual process of cell division, as well as steps preparatory to it. Here the genetic material of the cell is inactive: structural genes do not function.

This is the reason for the importance of studying the spatial organization in this interphase period: It is here that the fundamental, vitally important sysnthesis functions of the cell are determined.

The "inactively arranged" nucleus lends itself immediately to direct observation, since the chromosomes are contracted (in spiral form) and are able to absorb specific stains which bring out their outlines clearly.

With present methods it is impossible to observe the intact, three-dimensional structure of the interphase nucleus. Cytologists study the internal structure of the cell in a "two-dimensional" form, using what are called *stressed preparations*. In even the best of these preparations we can only view the "fragments" of intra- and interchromosomal connections. By comparing such two-dimensional pictures it is hoped to be able to reconstruct the spatial arrangement of the chromosomes in the interphase nucleus. The detection of this structure helps us understand the meaning of the genetic processes described previously, as well as their interrelations.

Two extreme variants of spatial organization in the chromosome set can be given a priori:

1) When the nature of the contacts (both within and between chromosomes) is random, so that in different cells of the same type there are different structures;

2) When the nature of the contacts specifically determines the spatial organization uniquely for all cells of a given type (given tissue).

Information about the spatial organization of a chromosome can be obtained with the help of the matrix of paired contacts of its various regions (Fig. 31). This particular matrix contains the reduced results

	102F	101F	100A	98C	94D	94A	89E	86D	84D	84A	83D	81F	80ABC	75C	73F/74A	71C	70C	70AB	67D	67A	65B	64C	61A	60F	59D
102F	1	1									1	1	1												
101F		1										1	1	1											
100A			1	1				1													1	1			
98C				1	1	1		1									1				1	1		1	
94D					1	1										1	1								
94A						1										1	1					1			
89E							1	1											1		1		1	1	
86D								1	1	1	1						1		1		1		1		
84D									1	1	1														
84A										1	1														1
83D											1	1				1							1		
81F												1										1			
80ABC													1												
75C														1	1	1	1	1	1	1	1		1		
73F/74A															1		1	1							
71C																1	1	1				1			
70C																	1		1	1		1			
70AB																		1	1			1			
67D																			1	1	1	1			
67A																				1	1	1			
65B																					1	1			
64C																						1	1	1	1
61A																							1	1	
60F																								1	
59D																									1
57AB																									
56F																									
42B																									
39E																									
39A																									
36D																									
35F																									
35A																									
33A																									
25E																									
25A																									
22AB																									
21A																									
19E																									
18F/19A																									
16D																									
12EF																									
11A																									
9AB																									
8B																									
7ABC																									
4C																									
3C																									
1A																									

	57AB	56F	42B	39E	39A	36D	35F	35A	33A	25E	25A	22AB	21A	19E	18F/19A	16D	12EF	11A	9AB	8B	7ABC	4C	3C	1A
102F														1								1		
101F			1	1										1									1	
100A		1													1			1						
98C		1																						
94D			1																1					
94A						1																		
89E							1		1				1											
86D			1									1					1							
84D						1					1													
84A																							1	
83D			1																					
81F																	1							
80ABC		1	1																			1		
75C						1	1									1							1	
73F/74A								1	1															
71C	1		1																		1	1		
70C														1			1							
70AB														1				1						
67D									1									1	1		1			
67A																								
65B																								
64C		1					1						1				1			1				
61A													1											1
60F	1	1								1		1	1									1	1	1
59D	1											1												
57AB	1	1																				1		
56F		1										1												
42B			1				1																	
39E				1		1	1																	
39A					1	1															1			
36D						1	1	1	1	1					1					1				
35F							1	1	1		1				1	1								
35A								1	1		1					1	1				1			
33A									1					1							1			
25E										1	1	1	1							1				
25A											1	1				1				1				
22AB												1				1	1						1	
21A													1		1									1
19E														1		1	1	1				1		
18F/19A															1	1	1				1			
16D																1	1	1		1				
12EF																	1	1		1	1		1	
11A																		1	1		1	1	1	
9AB																			1		1	1	1	
8B																				1	1		1	
7ABC																					1	1	1	1
4C																						1	1	1
3C																							1	1
1A																								1

Figure 31

obtained by a phase-contrast microscope from approximately 170 investigations of interphase cell nuclei from the salivary gland of *Drosophila*[†] [11].

[†]This type of cell is of particular interest to cytologists, since its chromosomes are *polynemic*, containing up to 1000 strands of DNA in cross-section. All these strands arise from an initial one through repetitive replication, without the subsequent separation of the daughter molecules. Thus these chromosomes are much thicker than usual, and are more easily stained and viewed in the interphase.

Of roughly a hundred of the chromosome segments (which were separated cytologically), only those 49 which participated in at least one contact are given in the table.

It must be said that the numbers contained in the initial experimental table specified the number of times the corresponding regions were observed to be in contact. Neighboring sections within individual chromosomes were found to be more frequently in contact. This is probably an artifact connected with the method of preparing the cells: In the destruction of the nucleus evidently the most easily destroyed connections are those between more distant regions of the chromosomes. The sample was sufficiently large that we may assume the basic contacts are determined. Therefore the initial matrix may be reformulated by replacing all the non-null elements with ones (see Fig. 31). A one means that the corresponding sections were observed to be in contact at least once.

To reconstruct the general picture of interchromosomal contacts we must partition the chosen regions into groups according to the similarities of their contacts and search for the structure of the connections among the groups.

If the spatial structure of the genes is random then the connections among the selected sections of the chromosome will surely be arranged in a complex form and will have a very complex macrostructure. On the other hand, if the spatial structure is fixed, the observed collections must consist of relatively isolated groups, corresponding to key knots in the structure.

The results of applying the structure algorithm support the latter supposition. The bounds for the degree of partition were set quite broad. The resulting graph has a complex form if the number m is greater than 11. However, beginning with 11 classes, practically all the regions of the chromosome were united into groups strongly connected within themselves and having no connections with each other (Fig. 32 and Table 7). Subsequent reductions in the number of classes had practically no effect on the form of the graph. Thus the given system actually represents a collection of isolated classes. However, one cannot suppose that the sections of the chromosome falling into one class necessarily form a single knot. We suppose that within the bounds of a particular group there may be several possible conjugations, i.e. in one nucleus A and B are in contact; in another, B and C; and in a third, A and C; but the three do not form a knot, and the real structure has the form A-B, C, or A-C, B, or B-C, A. In such a situation A, B and C may fall into one class as a result of classification.

An important feature of this graph is the presence of a connection between classes 3 and 6 in the absence of intraclass connections, and the presence of a connection between classes 8 and 9 in the absence of internal connections in class 8 (see Fig. 31). Apparently, this fact points out the method of formation of knots in the structure. We examine the simplest example of such a situation. Suppose that sections A, B and C form a knot, with the property that B is always connected

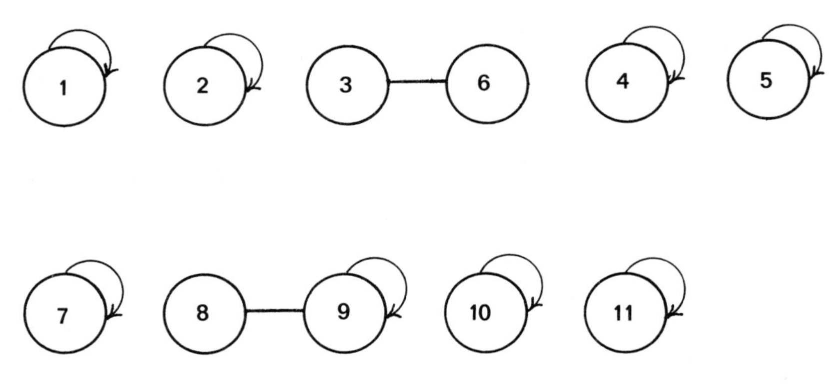

__Figure 32__. Classes of *Drosophila* chromosome regions according to the similarities of their non-homologous contacts (the result of a partitioning into 11 classes).

with A and C, but A and C are never in contact (A-B-C). Structures of this type have been encountered frequently in analysis. Then A and C are similar to one another, since they are both connected to B, though they do not contact each other. If a partition were given into two classes, then the graph would have this form:

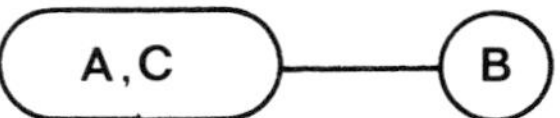

Consequently, graphs of the form described reflect the formation of a knot around some center, in which the sections associated with the center do not directly interact with one another. Thus the structure algorithm permits, in this case, not only the determination of the number of knots and the arrangement of their connections, but also the regularity of interaction of components within a knot.

So the results of analyzing the structure of a contact matrix confirm the presence of a fixed spatial structure in a set of interphase chromosomes.

Recombinations probably occur more frequently within the knots found, since recombinations imply spatial nearness of exchanged genetic sections. The ends of sections involved in chromosomal rearrangements (for example, inversions) must also lie in regions which form knots, since rearrangements arise from recombinations. A further, most important characteristic of these sections is the hypothesis of their later replication, if only the spatial structure of the nucleus is maintained up to the very beginning of nuclear division, since for replication of DNA a linear unwinding of the corresponding chromosome segments is necessary.

TABLE 7

The Contents of the classes corresponding to Fig. 32

1	102F, 101F, 80ABC, 42B
2	100A, 98C, 86D, 65B, 64C
3	94D, 94A, 75C, 73F/74A, 71C, 67D, 67A, 19E,
4	89E, 61A, 60F, 21A, 1A
5	84D, 84A, 83D, 81F
6	70C, 70AB
7	59D, 57AB, 56F, 22AB
8	39E, 39A, 35A, 33A, 18F/19A
9	36D, 35F
10	25EF, 25A, 16D, 12EF, 8B
11	11A, 9AB, 7ABC, 4C, 3C

It turns out that all of these conjectured events actually occur [11]. In the early papers of N. P. Dubinin and V. V. Khvostova [6] it is conjectured that complex chromosomal rearrangements (for example, inversions), which have arisen at the same moment, are the result of recombination exchanges in knots formed by segments which may belong to the same or to different chromosomes. The analysis of matrices constituted on the basis of the distribution of the ends of restructurings would, of course, permit more exact building of the knot structure of the interphase nucleus. However, the data necessary for this are not yet available [11].

Moreover, other, previously unexplained facts are in agreement with this structure, such as the change in the intensity of the recombination process in certain sections of chromosomes under the influence of inversions, occurring in other chromosomes: The majority of the corresponding sections fall in some one of the constructed classes (knots) of the structure.

This structure also partially clears up the following phenomenon, usual for higher organisms. With some exceptions (with which, in all probability, the locus *scute* is associated), genes which constitute a single functional system "working" on a single autonomous trait (for example, biochemical), are frequently located in different chromosomes or are far from one another within the bounds of one chromosome. Can it be that these genes are neighbors in the spatial structure of the interphase nucleus?

These systems have not been studied sufficiently to permit a final answer. However, in *Drosophila*, for a number of known gene systems (a group of genes controlling the development of pigments, a group controlling the structure of various fractions of ribosomal RNA, and others) this is exactly the situation [11].

Thus this example was included in the chapter not just as an application of a formal device: The spatial structure of the interphase nucleus may determine essential features of the functioning of complex systems of "gene-enzymes" in higher organisms.

The whole situation is reminiscent of the interallelic complementation problem discussed in the previous section. In both of these situations the initial material is represented by linear structures (polypeptide chains and chromosomes), but the realization of their inherent functional possibilities occurs after spatial packing, ensuring the proximity of sections responsible for a single function. The thought automatically presents itself that here we have an example in which nature uses the same principles in ensuring the functioning of the genetic system at different levels. At the same time, in the case of interallelic complementation the constructed graphs characterize exactly those semantic connections of the most important functional sections of the protein macromolecule. But it is still early to talk of constructing a proper model of the protein's structure, based on this information. On the other hand, in the second case we are able to show the key knots in the spatial organization of the chromosome as a whole, and on the basis of this are able to attack the description of the semantic features of the system. From the perspective of knowledge of the spatial arrangement of genetic systems in the interphase nucleus one may be led to understand the principles of functional coordination of separate blocks of the genetic program of cells in the process of their individual growth (i.e. ontogenesis).

Chapter 3. Graphs in the Analysis of Gene Evolution

§1. Trees and phylogenetic trees

1. The notion of a phylogenetic tree. In the preceding chapters we examined the uses of graphs for analyzing the structure and semantics of "contemporary" genetic texts. However, graph-theoretic methods can also be used for investigating the evolution of these texts (both structural and semantic).

The recent growth of interest in the evolution of biological macromolecules (nucleic acids and proteins) is completely natural. Since molecular events at the level of separate genes (mutations, recombinations etc.) form the basis for evolution in living organisms, the direct study of the former contributes significantly to the understanding of the evolutionary process as a whole. In comparison with classical investigations of evolution in living things, which are based on the comparative analysis of "external" morphological features, the study of evolution at the molecular level relies, as we shall see below, on incomparably more exact initial data. It is well to bear in mind also that changes in morphological features depend, in the final analysis, on elementary changes in genetic texts. It is not without reason that the coding biopolymers (DNA, RNA and proteins) are known as "the molecular documents of evolution" [99].

For natural reasons any attempts to describe the rise and evolution of genetic systems can be based only on the analysis of contemporary genetic texts. It is in the structure of these texts that we will seek traces of their descent. But the significance of such analysis is not exhausted by this. The modeling of the evolutionary processes of genetic systems helps clarify the causes, methods and principles of structural and functional organization in presently existing systems. In what follows we will illustrate a concrete definition of the hackneyed truth that "studying the past, one finds better understanding of the present." This will be accomplished by analyzing the evolutionary path of quaternary protein organization in the family of globins.

The following statement will serve as the basic supporting premise for the study of evolution in genetic systems: Similarity of genetic texts implies their common ancestry. Actually, the potential number of variants of a genetic text of average length (say a hundred to a thousand characters: amino acids and nucleotides) is astronomical. However, a simple calculation will show [30,31,52], that such diversity could not be selected out in a reasonable evolutionary period. Therefore, the independent appearance of similar texts by evolutionary means is extremely improbable. In other words, in nature the chances of randomly meeting two or more identical variants of genetic information is insignificant.

Thus, if two genetic texts contain noticeably similar features this is not by accident, it clearly must come about by divergent evolution of the texts (See page 162). As far as the process of convergence is concerned, it is realized only in very short fragments of genetic text[31].

The beginning point for the investigation of evolution in texts is quantitative estimation of the similarity or difference between biopolymers taken from different species, but which are synonymous in the sense that they fulfill the same molecular function. The distance matrix obtained for the species being considered is used to construct the *phylogenetic tree*, the distinctive genealogy of the given family of synonymous biopolymers, which reflects the path of its, primarily divergent, evolution.

Nucleotide texts have been seldom used for this purpose, since the corresponding molecular data are fragmentary. More frequently, in order to analyze the differences in genetic information the investigator has turned to the secondary data of amino-acid sequences (the primary structure of the polypeptide language).

The comparison of synonymous polypeptide or polynucleotide sequences can be performed by counting the number of symbol-by-smybol differences in homologous positions. The fixing of the fact of coincidence or non-coincidence of symbols in corresponding positions in two genetic texts assumes that replacements of symbols one for another are equally likely. This is reasonable for comparison of nucleotide texts, but for protein sequences it may be applicable only over quite long evolutionary time intervals. Over relatively short time intervals a more adequate hypothesis is the equi-probability of replacements of corresponding nucleotide triplets, which, due to the non-uniform degeneracy of the genetic code, leads to unequal probabilities of replacements of amino acids. As a first approximation this situation is handled by the so-called *minimal mutational distance*: the minimal number of nucleotide replacements required to pass from the codon of one of the amino acids being compared to the codon of the other. The matrix of minimal mutational distances for all 20 amino acids follows:

	1	2	3	4	5	6	7	8	9	10	11	12	13	14	15	16	17	18	19	20
1. Asp	0	2	2	2	1	1	2	1	3	1	1	2	2	2	2	3	2	1	2	1
2. Cys	2	0	2	1	3	2	3	2	3	2	1	2	2	3	3	3	2	1	2	1
3. Thr	2	2	0	2	2	2	1	1	1	1	2	1	2	1	1	2	2	2	1	2
4. Phe	2	1	2	0	3	2	3	2	2	2	1	2	2	2	1	2	1	1	1	2
5. Glu	1	3	2	3	0	2	1	1	2	2	2	2	1	2	2	2	2	1	3	1
6. His	1	2	2	2	2	0	2	2	3	1	1	1	1	1	2	3	1	2	2	2
7. Lys	2	3	1	3	1	2	0	2	1	1	2	2	1	1	2	2	2	2	2	2
8. Ala	1	2	1	2	1	2	2	0	2	2	2	1	2	2	1	2	2	1	2	1
9. Met	3	3	1	2	2	3	1	2	0	2	3	2	2	1	2	2	1	1	1	2
10. Asn	1	2	1	2	2	1	1	2	2	0	1	2	2	2	1	3	2	2	1	2
11. Tyr	1	1	2	1	2	1	2	2	3	1	0	2	1	2	1	2	2	2	2	2
12. Pro	2	2	1	2	2	1	2	1	2	2	2	0	1	1	1	2	1	2	2	2
13. Gln	2	2	2	2	1	1	1	2	2	2	1	1	0	1	1	1	1	2	3	2
14. Arg	2	1	1	2	2	1	1	2	1	2	2	1	1	0	1	1	1	2	2	1
15. Ser	2	1	1	1	2	2	2	1	2	1	1	1	1	1	0	1	1	2	1	1
16. Trp	3	1	2	2	2	3	2	2	2	3	2	3	1	1	1	0	1	2	3	1
17. Leu	2	2	2	1	2	1	2	2	1	2	2	1	1	1	1	1	0	1	1	1
18. Val	1	1	2	1	1	2	2	1	1	2	2	2	2	2	2	2	1	0	1	1
19. Ile	2	2	1	1	3	2	2	2	1	1	2	2	3	2	1	3	1	1	0	2
20. Gly	1	1	2	2	1	2	2	1	2	2	2	2	2	1	1	1	1	1	2	0

The common distance between two polypeptides is obtained by summing the minimal mutational distances for all positions of their primary structure.

It should be said that this distance measure does not begin to take full account of the diversity of mutational differences between polypeptides. In the real process of evolution the principle of the minimality of mutational transitions is scarcely observed in all cases. Besides individual replacements, other kinds of changes dissimilar to them occur, such as the deletion and insertion of nucleotides, as witness the fact that the lengths of a number of synonymous polypeptides are not identical†. Therefore the connection of mutational distances with real (paleontologic) times requires some corrections (the corresponding methods are described in [99,31]).

Before proceeding to the precise mathematical formulation of problems associated with constructing phylogenetic trees of proteins, we remark on the appropriateness of considering evolution not only in its syntactic (structural) aspects, as is usually done, but also in its semantic (functional) aspect, since the uniformities of action of natural selection are connected in the first place with the estimation of the

†To estimate the distance between two polypeptides in this situation, a mutual positioning is sought, for which the number of positions containing identical amino acids is maximized. Discontinuities in the texts are allowed in the process [62].

semantics of genes.

To make this more clear we note that the distance measure defined above reflects only the structural aspect: in calculating the distance between polypeptides all positions are considered to have identical weights. This corresponds to the mutational mechanism, which has no connection with the meaning of the information; so that a mutational replacement may affect any nucleotide of the cistron with equal probability. However it is clear that the fixation of a mutation in a population depends on the position in the primary structure where it arises. Mutations affecting functional centers (the semantic parts of the molecule), are, of course, controlled by more strict selection. Therefore, in calculating the "semantic" distance between proteins, positions within functional centers must be considered to have significantly greater weights than the others.

2. The metric generated by a tree. We will consider *weighted trees*, i.e. trees in which each edge ij has an associated numeric "length" $f_{ij}>0$. In such a tree it is natural to introduce a distance measure $d(i,j)$ between any two vertices as the sum of the lengths of the edges of the unique chain $[i,j]$, which joins them. This distance measure is clearly symmetric, and $d(i,i)$ is 0 by definition. Moreover $d(i,j)$ satisfies the triangle inequality $d(i,j) \leqslant d(i,k) + d(k,j)$.

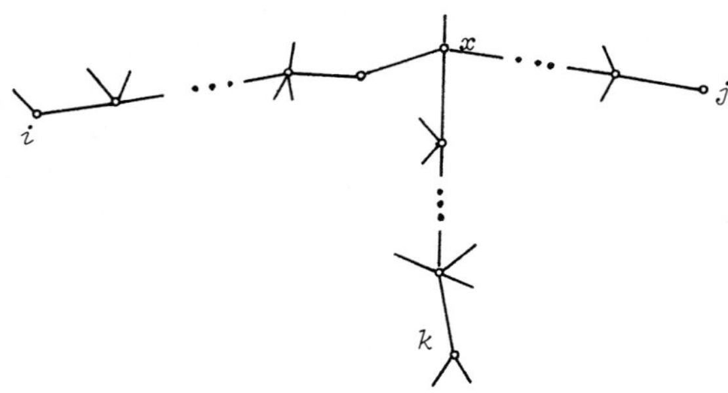

Figure 33

This is evident from examination of Fig. 33, which shows the mutual disposition of any three vertices of a tree. Here x is the first common vertex of the chains $[i,j]$ and $[k,j]$. Clearly, $d(i,k) + d(k,j) = d(i,j) + 2d(k,x)$, so that $d(i,j) \leqslant d(i,k) + d(k,j)$, equality being obtained if and only if $d(k,x) = 0$, i.e. $k \varepsilon [i,j]$.

From this it also follows that $d(k,x) = \frac{1}{2}[d(k,i)+d(k,j)-d(i,j)]$. Thus a weighted tree gives rise to the metric $d(i,j)$. In particular the system of distances $\|d(i,j)\|_1^N$ between pendant vertices of a tree is a metric.

A vertex of a tree which is adjacent to only one other vertex is said to be *pendant*, one adjacent to two vertices is *communicating*, and one adjacent to three or more vertices is a *branching* vertex. Sometimes non-pendant vertices are said to be *internal*.

The question arises, how completely is the initial tree characterized by its generated matrix $\|d(i,j)\|$ of distances between pendant vertices?

It is known that the metric $\|d(i,j)\|_1^N$ carries within itself information about all of the branching vertices: For, with each branching vertex x one can associate three pendant points i,j,k such that x is the first common vertex of the chains $[i,k]$ and $[j,k]$, with the distances $d(i,x)$, $d(j,x)$, and $d(k,x)$ determinable on the basis only of the given distances $d(i,j)$, $d(i,k)$, and $d(j,k)$, as was already done for $d(k,x)$. Therefore, in every tree which gives rise to the metric $d(i,j)$ of distances between pendant vertices, for these i,j and k there must exist a vertex y, distant from i,j and k by the same amounts as x, and also being the first common vertex of the chains $[i,k]$ and $[j,k]$, and in particular, a branch point.

At the same time, communicating vertices are not uniquely determined by the system $\|d(i,j)\|_1^N$. The matrix $\|d(i,j)\|$ is unchanged by deletion of a communicating vertex x, if the two adjacent edges ax and xb are replaced by a single edge ab with length $f_{ab} = f_{ax} + f_{xb}$. Similarly, any edge ab can be divided into two successive adjacent edges ay and yb with a new communicating vertex y, by defining f_{ay} and f_{yb} arbitrarily, subject to the restriction that

$$f_{ab} = f_{ay} + f_{yb}.$$

Thus a tree realizing the matrix $\|d(i,j)\|_1^N$ (as its matrix of distances between pendant vertices) is determined to within the existence of communicating vertices. We formulate a more complete and exact statement.

Theorem 1. A symmetric matrix $\|d_{ij}\|_1^N$ with zero elements on the main diagonal is realized by a tree with N pendant vertices such that $d(i,j) = d_{ij}$ $(i,j=1,\ldots,N)$ if and only if for any $i,j,k,l = 1,\ldots,N$

(1) $d_{ij} < d_{ik} + d_{kj}$;

(2) among the numbers $d_{ij}+d_{kl}$, $d_{ik}+d_{jl}$, and $d_{il}+d_{kj}$ two are identical, and the third does not exceed them.

A tree with the minimal number of edges, which realizes a matrix $\|d_{ij}\|$, is uniquely determined (to within an isomorphism which preserves the numbering of the pendant vertices).

Proof. If L is a tree for which $d(i,j) = d_{ij}$ then condition (1) means that no pendant point belongs to a chain which joins two other pendant points.

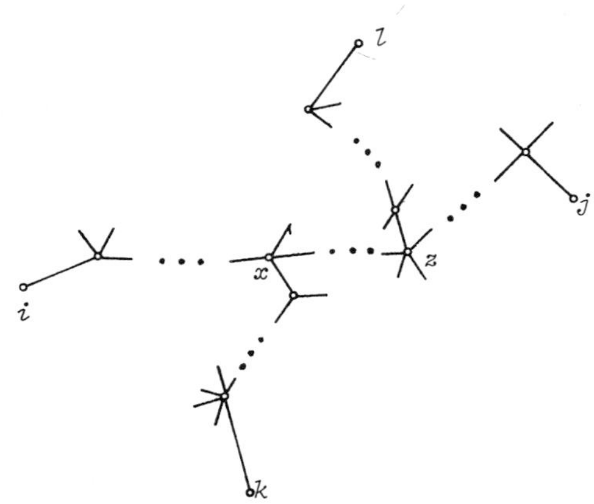

Figure 34

The meaning of condition (2) is clear from Fig. 34, where a system of chains is shown which joins the points i, j, k and l. Here x is the last common point of the chains $[i,j]$ and $[i,k]$, and z is the last common point of the chains $[l,j]$, $[l,i]$ and $[l,k]$. It is possible that $x = z$. From the indexing shown we have

$$d(i,j)+d(k,l)=d(i,l)+d(k,j) \geqslant$$

$$\geqslant d(i,k)+d(l,j).$$

Equality of all three quantities is obtained when $x = z$.

We now prove the reverse assertion: if (1) and (2) hold then there exists a tree having the distances $\|d_{ij}\|$, with its minimal (with respect to the number of its edges) realization being determined to within an isomorphism.

The proof will proceed by induction on N. For $N=2$ conditions (1) and (2) are trivially satisfied, and the matrix $\|d_{ij}\|$ reduces to the single number d_{12}, so that the tree sought is formed by the edge 12 with length d_{12}, with its minimal realization uniquely determined.

Suppose that the statement of the theorem is true for N pendant vertices. We will show that it is true for $N+1$ pendant vertices also.

We construct the tree L_N corresponding to the first N vertices. Consider now the numbers $d_{N+1,i} + d_{N+1,j} - d_{ij}$ $(i,j=1,\dots,N)$, which are positive, from (1). Now we fix the pair (i,j) for which $d_{N+1,i} + d_{N+1,j} - d_{ij}$ is minimal, and consider the three numbers

$$d'_{N+1} = \frac{1}{2}\left(d_{N+1,i}+d_{N+1,j}-d_{ij}\right),$$

$$d'_i = d_{N+1,i} - d_{N+1} = \frac{1}{2}\left(d_{ij} + d_{N+1,i} - d_{N+1,j}\right),$$

$$d'_j = d_{N+1,j} - d_{N+1} = \frac{1}{2}\left(d_{ij} + d_{N+1,j} - d_{N+1,i}\right),$$

which are positive, from (1).

In the chain $[i,j]$ of the tree L_N we introduce a new vertex x (by dividing one of the edges into two), in such a way that $d(x,i) = d'_i$ and $d(x,j) = d'_j$ (If in chain $[i,j]$ there is already a vertex x with $d(x,i) = d'_i$ and $d(x,j) = d'_j$ then it is not necessary to introduce a new one). Clearly, such an introduction is correct, since $d(i,x) + d(x,j) = d'_i + d'_j = d_{ij}$ by definition. The tree L_{N+1} is now obtained by adjoining a new pendant vertex $N+1$ to the vertex x by means of an edge whose length is d'_{N+1}.

We will show that in the tree thus constructed $d_{N+1,l} = d(N+1,l)$ for all $l = 1, \ldots, N$. It is clear that

$$d(N+1,i) = d(N+1,x) + d(x,i) = d'_{N+1} + d'_i = d_{N+1,i}$$

$$d(N+1,j) = d(N+1,x) + d(x,j) = d'_{N+1} + d'_j = d_{N+1,j}.$$

Now suppose $l \neq i,j$. By the selection of i and j we have:

$$d_{i,N+1} + d_{j,N+1} - d_{ij} \leqslant d_{l,N+1} + d_{j,N+1} - d_{lj},$$

so that $d_{lj} + d_{i,N+1} \leqslant d_{l,N+1} + d_{ij}$.

We now consider the chains $[l,i]$, $[l,j]$ and $[l,N+1]$. Let k be the last common vertex of these chains. By the construction it belongs to either $[i,x]$ or $[j,x]$, since $[N+1,x]$ consists of a single edge. Suppose, say, that k belongs to $[i,x]$, so that

$$d(l,N+1) + d(i,j) = d(i,N+1) + d(l,j).$$

Then $d(i,k) \leqslant d(i,x)$, so that

$$2d(i,k) = d_{ij} + d_{il} - d_{jl} \leqslant d_{ij} + d_{i,N+1} - d_{j,N+1} = 2d(i,x).$$

Hence

$$d_{il} + d_{j,N+1} \leqslant d_{jl} + d_{i,N+1}.$$

Considering the preceding inequality, we obtain

$$d_{il} + d_{j,N+1} \leqslant d_{jl} + d_{i,N+1} \leqslant d_{l,N+1} + d_{ij},$$

from which, by condition (2) of the theorem

$$d_{jl} + d_{i,N+1} = d_{l,N+1} + d_{ij}.$$

Hence it follows that

$$d(l,N+1) = d(i,N+1) + d(l,j) - d(i,j) = d_{i,N+1} + d_{lj} - d_{ij} = d_{l,N+1}.$$

Q.E.D.

It remains to show that the tree thus obtained is minimal and is determined to within an isomorphism, which preserves the numbering of pendant vertices. We obtained L_{N+1} from L_N by the addition of no more than two vertices: one of these was the vertex $N+1$, and the other, x, was added to the chain $[i,j]$ with $d(x,N+1) = d_{N+1}$, which had to be contained in L_{N+1} as the last common point of the chains $[N+1,i]$, $[N+1,j]$, in order that the combination $d(N+1,i) + d(N+1,j) - d(i,j)$ have a "fixed" meaning. This proves both the minimality and the uniqueness of L_{N+1}, Q.E.D.

Theorem 1 is a modification of a theorem of Smolenski and Zaretski [7] in the case when the tree is weighted. For results on the representation of metrics by minimal graphs (with respect to the number of edges) of a common form see [8].

We give several corollaries of Theorem 1.

Corollary 1 [93]. A metric $\|d_{ij}\|_1^N$ is generated by a tree if and only if to each subset of four objects their corresponds a submetric which characterizes the tree.

Actually, conditions (1) and (2) of the theorem are formulated for four-element subsets of the initial set, so that it is necessary and sufficient to check their validity for just such subsets.

The next corollary is connected with a notion of considerable importance to us: that of a dendrogram. A *dendrogram* (*tree*) is a rooted tree in which each internal vertex has exactly two edges leading from it to the next level (and one edge leading to it from the preceding level). Thus in a dendrogram all internal vertices besides the root have three "neighbors", while the root is a communicating vertex. The importance of this notion is connected with one of the basic heuristic rules of the theory of macromolecule evolution: in the process of evolution each "antecedent" biopolymer species *diverges* (divides) into two species of "descendants". Therefore the desired phylogenetic tree is in the form of a dendrogram. In fact, divergent evolution need not be dichotomous (See page 160).

Corollary 2. The metric $\|d_{ij}\|_1^N$ is realized by a dendrogram if and only if conditions (1) and (2) of Theorem 1 are satisfied, as well as the following condition (3): for any i,j,k,l, among the numbers $d_{ij}+d_{kl}$, $d_{ik}+d_{jl}$, and $d_{il}+d_{jk}$ there must be differences.

Actually, analysis of Fig. 34 shows that if some vertex of the tree is incident on more than three vertices, then taking as the four vertices i,j,k,l those for which the paths $[x,i]$, $[x,j]$, $[x,k]$ and $[x,l]$ pass through different edges incident on x, we obtain the equality of the mentioned numbers.

We now consider a metric of a special form, for which the construction of a tree is an essentially trivial procedure. The metric $\|d_{ij}\|$ is an *ultrametric* (sometimes called a *Berovski metric* [47]), if it satisfies the stronger inequality: for any i,j,k

$$d_{ij} \leqslant \max(d_{ik}, d_{kj}).$$

This inequality means that out of the three numbers d_{ij}, d_{ik} and d_{kj}, two coincide, and the third does not exceed their common value. From this it follows that an ultrametric automatically satisfies condition (2) of Theorem 1. We will not prove this fact immediately, rather, we will first make clear the structure of the matrix $\|d_{ij}\|$ for an ultrametric.

Consider the indices i, j for which the value d_{ij} is minimal ($i\neq j$). For any $k\neq i,j$, $d_{ik}=d_{kj}$, since all three numbers d_{ij}, d_{ik} and d_{jk} cannot be pair-wise distinct. Consider further the *a-similarity graph* for $a = d_{ij}$, i.e. that graph G_a, the vertices of which are objects, and which has edges (k,l) only if $d_{kl}\leqslant a$. From what has been said the relation of d_{ij}-similarity is an equivalence: if $(k,l)\varepsilon G_a$ and $(l,m)\varepsilon G_a$, then $d_{kl} = d_{ij} = d_{lm}$ because of the minimality of d_{ij}. But then $d_{km} = d_{ij}$ by definition of the ultrametric, i.e. $(k,m)\varepsilon G_a$.

The classes of the corresponding partition $R = \{R_1, \ldots, R_m\}$ are "similar" not only in their internal makeup but also externally, i.e. if $k,k'\varepsilon R_s$ and $l,l'\varepsilon R_t$ then $d_{kl}=d_{k'l'}$. If $l=l'$ or $k=k'$ then this equality was proved in the preceding discussion. Now suppose $l\neq l'$, $k\neq k'$; then since $d_{kk'}$ is minimal, $d_{kl}=d_{k'l}$. Similarly, $d_{lk'}=d_{l'k'}$ from the minimality of $d_{ll'}$. Comparing these relationships we obtain the equality $d_{kl}=d_{k'l'}$.

Thus the minimal distance d_{ij} along with the partition $R=\{R_1,\ldots,R_m\}$ for the d_{ij}-similarity graph with distances $\|D_{st}\|_1^m$ between classes, where $D_{st}=D_{kl}$ for $k\varepsilon R_s$, $l\varepsilon R_t$, completely defines the initial ultrametric. This measure $\|D_{st}\|_1^m$ is also clearly an ultrametric, and the preceding construction is applicable to it.

If we designate the various values taken on by d_{ij} $(i,j=1,\ldots,N)$ by $d_1<d_2<\ldots<d_n$, the d_i-similarity graphs $(i=1,\ldots,n)$ characterize the imbedded sequence of partitions R^i $(i=1,\ldots,n)$, $R^1\subseteq R^2\subseteq\ldots\subseteq R^n$, where R^n is the universal partition consisting of a single class. This system of partitions completely characterizes the rooted tree giving rise to the ultrametric $\|d_{ij}\|$.

The bottom-most (level 0) layer of the tree contains N pendant vertices, corresponding to the initial objects; vertices of layer k ($k=1,\ldots,n$) correspond to classes of the partition R^k; the vertex at layer n is the root. Any vertex i of layer k is directly connected with that vertex j of layer $k+1$ which is associated with the class of the partition R^{k+1} containing the class of R^k to which the given vertex belongs.

These classes may coincide, i.e. the vertices may be communicating vertices, and not necessarily branching ones. The length f_{ij} of an edge is defined by the formula

$$f_{ij} = \frac{d_{k+1} - d_k}{2}$$

(with $d_0=0$), so that the lengths of all the edges joining two neighboring layers are identical.

We will now show that this tree gives rise to the initial ultrametric. Suppose $d_{ij} = d_k$. This means that the pendant vertices i and j are "subordinate" to the common vertex x of layer k, which is the last common vertex of the chains joining the root with i and j. Each of the chains $[i,x]$ and $[j,x]$ contains k edges, with total length

$$\frac{d_1}{2} + \frac{d_2-d_1}{2} + \ldots + \frac{d_k-d_{k-1}}{2} = \frac{d_k}{2},$$

so that $d(i,j) = d_k = d_{ij}$, Q.E.D.

Generally speaking, the tree we have obtained is not minimal. The minimization of the number of vertices can be carried out by means of the procedure already described for deleting communicating vertices.

The following theorem has been proved in the process of this construction.

<u>Theorem 2</u>. A metric $\|d_{ij}\|_1^N$ is an ultrametric if and only if it is realized by a rooted tree (possibly having communicating vertices), all pendant vertices of which are at the bottom layer, and in which the lengths of the edges joining vertices of a given layer with those of another are determined only by the two layers and do not depend on specific vertices.

Loosely speaking, an ultrametric corresponds to a variant of protein evolution in which the divergent species accumulate essentially the same number of amino acid replacements in the same period of time, i.e. the speed of evolution of the protein sequences is constant for the various sections of the tree.

<u>Corollary 3</u>. An ultrametric d_{ij} corresponds to a dendrogram if and only if for any i,j and k ($i\neq j\neq k$) at least two of the numbers d_{ij}, d_{jk} and d_{ik} differ from one another.

Actually, the fact that in a rooted tree described by Theorem 2 some vertex is adjacent to three edges joining it with vertices of the next lower level, implies there are at least three pendant vertices i, j,k, "dominated" by this vertex, so that the distances among i, j and k are equal: $d_{ij} = d_{jk} = d_{ik}$.

3. <u>The construction of dendrograms</u>. Though Corollaries 2 and 3 characterize dendrograms, it is very seldom that real matrices satisfy their criteria. To construct the dendrogram for an arbitrary matrix in the literature, as a rule the following modified form of the unification algorithm is used.

At step $N-m$ the partition $R^m = \{R_1, \ldots, R_m\}$ and the matrix of average distances

$$d(R_s, R_t) = \frac{1}{N_s N_t} \sum_{i \in R_s} \sum_{j \in R_t} d_{ij}.$$

is examined. At the first step $(m=0)$ R^0 is the trivial partition into single-element classes with $d(R_s, R_t) = d_{st}$.

Sometimes, in place of the average distance between objects of classes, the median distance or maximum or minimum distance is used as the distance between classes.

An R_s and R_t are selected for which $d(R_s, R_t)$ is minimal $(s,t=1,\ldots,m)$ and then R_s and R_t are unified by the passage from an $m\times m$ matrix $\|d_{ij}\|$ to a new $(m-1)\times(m-1)$ matrix, replacing the two rows (and two columns) corresponding to R_s and R_t by one row (and column) corresponding to $R_s \cup R_t$ according to the formula:

$$d(R_s \cup R_t, R_v) = \frac{N_s d(R_s, R_v) + N_t d(R_t, R_v)}{N_s + N_t} \quad (v \neq s,t).$$

The dendrogram is determined by unifications of classes as branch points.

Consider for example the matrix

	1	2	3	4
1	-	21	8	26
2		-	27	45
3			-	20
4				-

At the first step the closest objects 1 and 3, with $d_{13}=8$ are unified, producing the 3×3 matrix of average distances:

	$\overline{1,3}$	2	4
$\overline{1,3}$	-	24	23
2		-	45
4			-

Then the classes {1,3} and {4}, and finally there remains only the unification of the two classes {1,3,4} and {2}. These successive unifications generate the following dendrogram (Fig. 35).

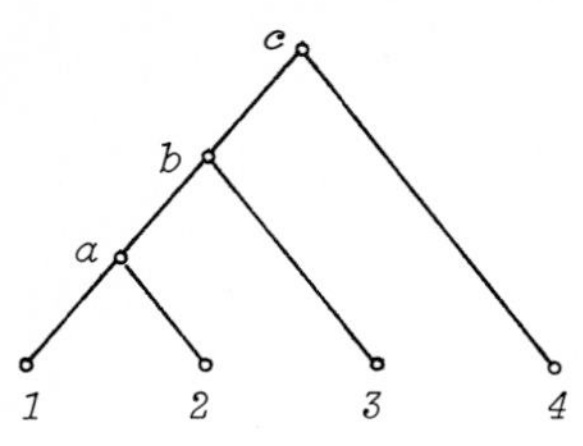

Figure 35

Now to determine edge lengths in this dendrogram we use the previously noted fact that in a tree the distance from any pendant vertex i to any "internal" vertex a can be uniquely determined by the indication of those vertices j and k, for which a is the last common vertex of the chains $[i,j]$ and $[i,k]$, according to the formula

$$d(i,a) = \frac{d(i,j)+d(i,k)-d(j,k)}{2},$$

Now to define the distance between arbitrary internal vertices a and b connected by the edge ab it is sufficient to take any pendant vertex i for which a is contained in the chain $[i,b]$, and define $d(a,b)=$ $=d(i,b)-d(i,a)$. However, the result generally depends on i, and therefore it is necessary to average the distance over all pendant i, bearing in mind that some i are connected with b through a and others with a through b. The following procedure is usually used [65,84].

To each internal vertex a (except the root) of the dendrogram there correspond three edges incident on it, and correspondingly, three sets of dependent vertices, according to which of these edges belongs to the chain joining the pendant vertex with a. These sets form a partition $R^a = \{R^a_1, R^a_2, R^a_3\}$ of the set of pendant vertices, which characterizes the vertex a. The root corresponds to a partition into two classes. For example, for the dendrogram of Fig. 35 $R^a = \{\{1\},\{3\},\{2,4\}\}$, $R^b =$ $= \{\{1,3\},\{4\},\{2\}\}$, and with the root is associated the partition into two classes: $R^c = \{\{1,3,4\},\{2\}\}$.

For a given vertex a let D_{st} $(s,t=1,2,3)$ designate the distance between the classes R^a_s and R^a_t of the partition R^a. The average distances from a to R^a_1, R^a_2, and R^a_3 are, respectively $D^a_1 = \frac{1}{2}(D_{12}+D_{13}-D_{23})$, $D^a_2 = \frac{1}{2}(D_{21}+D_{23}-D_{13})$, $D^a_3 = \frac{1}{2}(D_{31}+D_{32}-D_{12})$. If a is the root, then $D^a_1 = D^a_2 = \frac{D_{12}}{2}$.

The determination of edge lengths is now carried out in successive stages, beginning with some vertex of the lowest level.

For example, for the dendrogram of Fig. 35 we have

$$D^a_1 = \frac{8+(21+26)/2-(20+27)/2}{2} = 4,$$

$$D^a_2 = \frac{1}{2}(8+(20+27)/2-(21+26)/2 = 4,$$

$$D^a_3 = \frac{1}{2}((20+27)/2+(21+26)/2-8) = 19.5.$$

This fixes the lengths of edges $1a$ and $2a$ at 4. Similarly we obtain

$$D_1^b = 1, \qquad D_2^b = 22, \qquad D_3^b = 23.$$

This leads to $f_{ab} = D_a^b - \frac{1}{2}(D_1^a + D_2^a) = -3$, if we calculate from "below", and to the same result $f_{ab} = D_3^a - \frac{1}{2}(D_2^b + D_3^b) = -3$, if the calculation is done from "above." As is evident, this procedure may also lead to negative lines, if, as in the given case, distances are not in agreement with the tree topology. Similarly $f_{4b} = D_2^b = 22$.

For the root c

$$D_1^c = D_2^c = \tfrac{1}{2}\cdot 31 = 15.5.$$

Therefore from "below" $f_{bc} = D_1^c - \frac{1}{2}(D_1^b + D_2^b) = 4$, and from "above" $f_{bc} = D_1^c - D_2^c = 7.5$, so that "on the average" $f_{bc} = 5.7$. Now the correction

$$f_{2c} = D_3^b - f_{bc} = 23 - 5.7 = 17.3.$$

can be made.

Thus the heuristic procedure described leads to the edge lengths shown in Fig. 36, which give distances between pendant vertices only remotely resembling the initial distances.

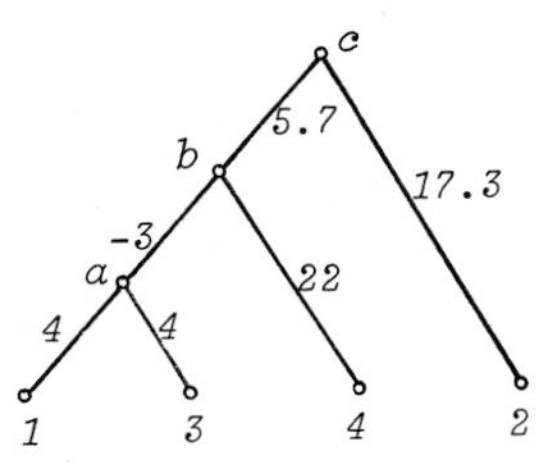

Figure 36

It must be added that the initial data are exactly realized by the tree of Fig. 37,a, which can be transformed into a dendrogram in many ways by the introduction of a new vertex-root between any two neighboring vertices (one of these being shown in Fig. 37,b).

This means that unification may give incorrect results even for a matrix realized by a tree.

We note that negative lengths may appear for edges of a tree not only from inadequacies of the unification algorithm being used, but also for the objective reason that at certain evolutionary levels different, divergent protein sequences may independently register identical amino acid replacements (parallel or *convergent* evolution).

To "correct" the results of the unification algorithm a number of indicators of correspondence between the initial tree and the dendrogram constructed from it have been proposed.

For example, the *Fitch-Margoliash coefficient* [65] measures the average relative squared error of the distances calculated for the dendrogram and the initial distances d_{ij} between objects:

$$\phi = \sqrt{\frac{1}{[N(N-1)/2]-1} \sum_{i<j} \left(\frac{d_{ij}-d(i,j)}{d_{ij}}\right)^2} . \tag{1}$$

Restructuring of a dendrogram for any tree is done in such a way as to minimize ϕ. The more simple, weighted sum of squared differences

$$\chi^2 = \sum_{i<j} \left(\frac{d_{ij}-d(i,j)}{d_{ij}}\right)^2 , \tag{2}$$

can be used for this since χ^2 has the same monotonic character as ϕ.

The meaning of χ^2 is a squared error estimate of the representation of the numbers d_{ij} by the distances $d(i,j)$, in which the (i,j)th term is smaller , the larger d_{ij} is, other things being equal. In other

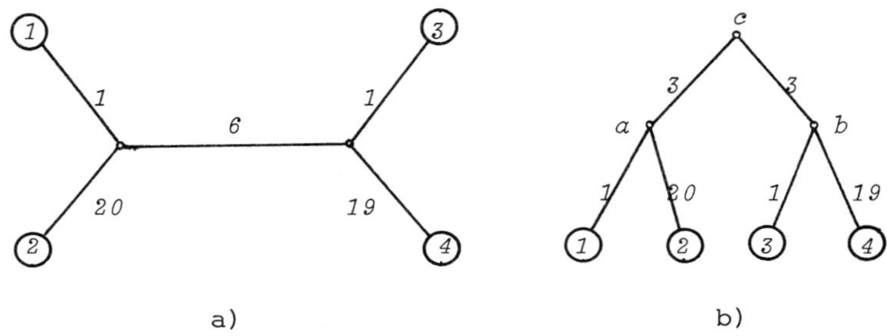

Figure 37

words, the error of approximation of small values of d_{ij} receives greater weight in (2) than that for larger values of d_{ij}. This seems very reasonable, since it corresponds to the accumulation of errors in the course of evolution.

Sums of squares such as (2) play the additional role of aids in constructing a tree for a matrix $\|d_{ij}\|$, and also in determining edge lengths in accord with the idea that the smaller the value of χ^2, the better is the approximation of the matrix $\|d_{ij}\|$. This idea, the method of least squares, was developed in the undergraduate work of A. Zharkiy (Novosibirsk State University, 1977).

The corresponding construction in terms of vectors is as follows.

Suppose a tree has N pendant vertices and M edges numbered 1 to M. The tree obtained from a dendrogram by the removal of the root (the only communicating vertex) has $M = 2N-3$ edges.

The topology of a tree is uniquely characterized by the $\frac{N(N-1)}{2} \times M$ matrices $B = \|b_{ij,k}\|$ $(i\ j,\ k=1,\ldots,M)$, where $b_{ij,k}$ equals $1/d_{ij}$, if the kth edge belongs to the unique chain connecting the ith and jth pendant vertices. Otherwise $b_{ij,k} = 0$. We designate by f_k the length of the kth edge, and the vector of edge lengths $(f_1,\ldots,f_M)$ by f. Then Then the product Bf is the $\frac{N(N-1)}{2}$-element vector of quantities $\frac{d(i,j)}{d_{ij}}$ $(i<j)$. Let $\mathbf{1}$ be an $\frac{N(N-1)}{2}$-element vector consisting of all ones. Then the indicator (2) may be expressed in the form of the scalar square

$$\chi^2 = (\mathbf{1}-Bf)\cdot(\mathbf{1}-Bf) \tag{3}$$

of the vector $\mathbf{1}-Bf$. Here, as usual, $x\cdot y$ is the scalar product $\sum_k x_k y_k$ of the vectors $x = (x_k)$ and $y = (y_k)$, such that $x\cdot x = \sum_k x_k^2$.

For a given topology of a tree, i.e. a matrix B, to find the vector $f = (f_1, \ldots, f_M)$ of edge lengths which minimizes the sum of squares (3), one must equate to zero the gradient of (3) as a function of f. In this case, as is usual in the method of least squares, f must be a solution of the system of equations

$$B^T\mathbf{1} = B^T Bf, \tag{4}$$

where B^T designates the transpose of B. Written in terms of vector elements (4) becomes

$$\sum_{i<j} b_{ij,k} = \sum_{l=1}^{M}\left(\sum_{i<j} b_{ij,k}b_{ij,l}\right)f_l \qquad (k=1.,,,.M). \tag{5}$$

Another indicator has been proposed by Moore [84]. In the case when a metric is realized in a tree the *Moore coefficient* for each internal vertex a of the derived dendrogram estimates the degree of deviation of the partition R^a from the collection of partitions corresponding to vertices of the "correct" tree.

For arbitrary objects i,j and k let

$$F(i,j,k) = \tfrac{1}{2}(d_{ij}+d_{ik}-d_{jk})$$

be the distance from i to the last common vertex of the chains $[k,i]$ and $[k,j]$, i.e. to the last common ancestor of species i and j. Then:

$$M(a) = \sum_{(i_1,i_2,i_3)} \sum_{i\in R_{i_1}} \sum_{j\in R_{i_2}} \left[\sum_{k\in R_{i_3}} F^2(i,j,k) - \frac{\left(\sum_{k\in R_{i_3}} F(i,j,k)\right)^2}{N_{i_3}}\right]. \tag{6}$$

Here the outer summation is made for all permutations of the classes of the partition R^a, and the term in square brackets measures the dispersion of the distribution of distances $F(i,j,k)$ from i to its last common ancestor with j (depending on k). Clearly, if $\|d_{ij}\|$ is realized by a dendrogram then $M(a)=0$ if and only if R^a is one of the partitions corresponding to the dendrogram [84]. This permits the use of $M(a)$ for directed improvement of the dendrogram for $M(a)>0$.

Consider the smallest neighborhood of a vertex a of the dendrogram given in Fig. 38,a. Here a_1 and a_2 are "offspring" of a, while a and b

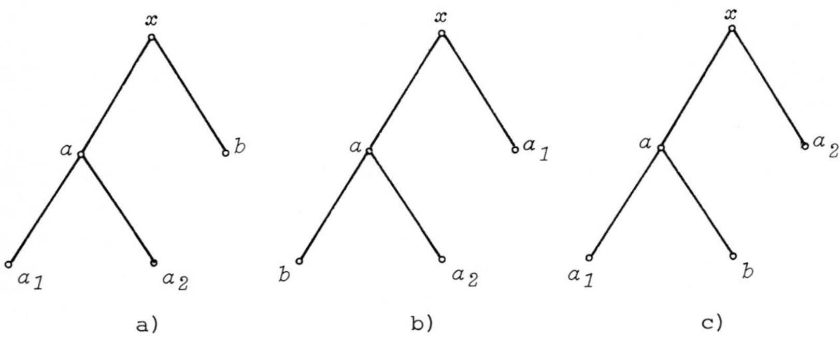

Figure 38

are "offspring" of x. Two methods can be proposed (Fig. 38,b,c) for modifying this local neighborhood while maintaining proper relationships of those parts of the dendrogram attached to a_1, a_2 and b.

An algorithm for successive improvement consists in changing one node at a time (if it decreases $M(a)$) for those nodes with $M(a)>0$. The sum $M = \sum_a M(a)$ acts as an overall indicator of the quality of a restructuring.

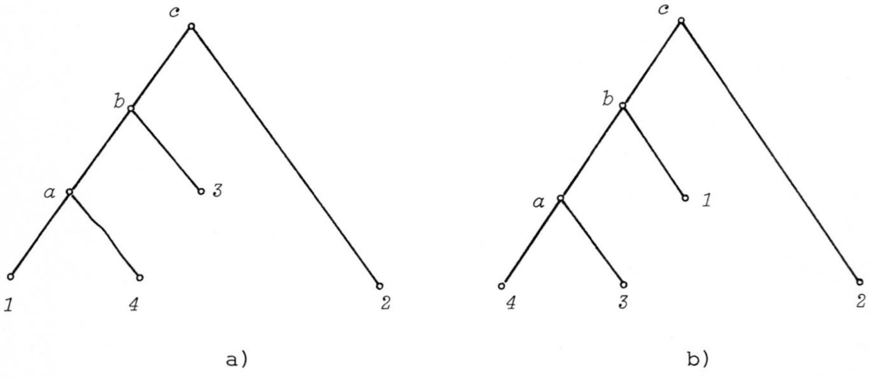

Figure 39

For the tree of Fig. 36 $M(a) = 36$, $M(b) = 18.5$. Consider the two restructurings of the neighborhood of vertex a (Fig. 39), where the Moore coefficient is maximized.

For the variant in Fig. 39,a we have $M(a) = 18$, $M(b) = 36$, so that the total error has not changed. For the dendrogram of Fig. 39,b $M(a)=$ $=M(b)=0$, showing that this dendrogram matches the initial matrix exactly.

Unfortunately the convergence of this algorithm to an exact solution has not been investigated.

Thus the question of the best approximation of a given metric by means of a tree still awaits solution. However in those cases when the speed of fixation of mutations is constant, so that one can expect $\|d_{ij}\|$ to be nearly an ultrametric, the described procedures provide completely "reasonable" solutions (See §2).

4. Reconstructing the probable structure of ancestral successions.

The detection of the facts of divergence of species in their order of occurrence by time is not the only use of such trees. For a given dendrogram one can also try to reconstruct the probable structure of the ancestral proteins, on the basis of information about the structure of contemporary protein species corresponding to the pendant vertices of the dendrogram.

For this purpose we make simplifying assumptions, corresponding, to one degree or another, with the real evolutionary process.

First, we assume that for all proteins under consideration one can establish the homologous positions in the primary structure which are the results of mutations having taken place in the same position of the ancestral protein. Second, these homologous positions will be considered independent from one another with respect to evolution, so that successive replacements in a single position (and, correspondingly, in the coding DNA triplet) are not functionally connected with those of other positions.

Thus the problem may be formulated for each separate position of the primary structure of a protein as follows: From information about the codons corresponding to pendant vertices of a dendrogram, reconstruct the codons associated with its internal vertices. In this process, "codon descendants" must resemble as much as possible their "ancestors."

We consider a more precise statment of the problem. Let X be the set of all vertices of a given dendrogram, with $A \subset X$ being the set of pendant vertices. Let Y designate the set of 61 semantically meaningful codons (without the "punctuation marks": UGA, UAA and UAG). With each pendant vertex $i \in A$ associate a set $F_i \subseteq Y$ of triplets, which code for the amino acid corresponding (in the position being considered) to vertex i.

A single-valued mapping $f: X \longrightarrow Y$ is *admissible* if for each $i \varepsilon A$ $f(i) \varepsilon F_i$.

The distance $d(y, y')$ between two codons y and y' is the number of corresponding positions in which they differ. The *length of the mapping* $f: X \longrightarrow Y$ is the total of the distances between $f(a)$ and $f(b)$ for all edges ab in the given dendrogram.

A *characteristic* mapping is a minimal length admissible mapping. The problem of constructing a characteristic mapping is a formalization of the problem of reconstructing ancestral protein sequences. We emphasize that this is a problem of data approximation, not of data representation

It is evident that if two pendant vertices i and j, adjacent to a "parent" vertex a, have identical amino acids, i.e. $F_i = F_j$, then for a characteristic mapping f $f(i) = f(j) = f(a)$. In fact, if $f(a)$ does not coincide with $f(i) = f(j)$, setting $f(a)$ equal to $f(i) = f(j)$ we obtain a reduction in the length of the dendrogram, independent of the extent to which the initial $f(a)$ differs from the codon of the third vertex adjacent to it. This property permits the replacement of two neighboring pendant vertices by their common ancestor, reducing in the process the tree being considered.

Still another possibility for reducing the calculations lies in the properties of synonymity of the genetic code. In many cases synonymous codons are obtained from one another by the replacement of one of their components. For example, the exchange in the third position of a codon of the component (nucleotide) U for C is always a synonymous transformation. From Table 1 (page 3) it is evident that under further restriction of the amino acids considered, still other possibilities appear for synonymous transformation by a reduction of the number of symbols in the jth component.

It is easy to see that if f is an admissible mapping, then a synonymous transformation in the jth component can only reduce the distance between any neighboring vertices. This means that we can produce at the beginning the synonymous transformations of the sets F_i, reducing their diversity, and then still seek a characteristic mapping.

We now describe a method for building a characteristic mapping. This method associates each internal vertex a with the set of those and only those codons which can be images of a under characteristic mappings. The method was proposed by Fitch [63] for the case when each pendant vertex is characterized by a single codon (all the F_i are single-element sets); and later in [85] it was shown that it is also suitable in the general case.

The construction is carried out in terms of the tree corresponding to the given dendrogram (i.e. obtained by excluding the root and uniting both its adjacent edges into one with their combined length). Fig. 40 shows the tree with given F_i, corresponding to the dendrogram of Fig. 35.

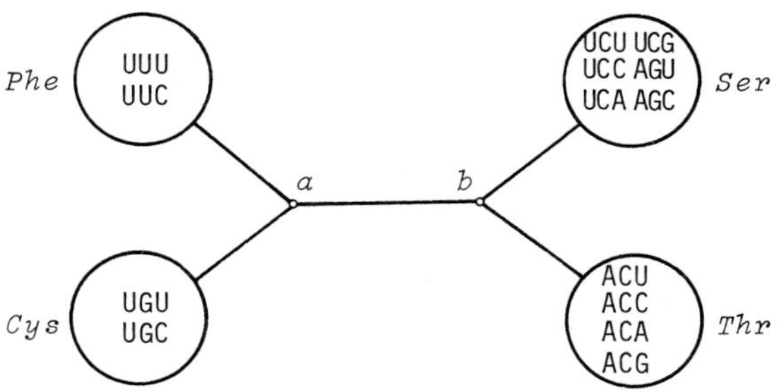

Figure 40

We first produce synonymous transformations of this tree by translating the final characters of all codons to U (in the given case this is possible); obtaining the tree of Fig. 41.

Now each pair *a*, *b* of adjacent vertices of the tree determines the mentioned sets $F(a,b)$ of codons, indexed by the numbers *0*, *1*, *2*. This construction is performed inductively, beginning with the pendant vertices. Each pendant vertex *i* is adjacent to a single internal vertex *a*; so we set $F(i,a) = F_i$, with the indices of all codons being set equal to zero. Each internal vertex *a* is adjacent to exactly three vertices *b*, *c* and *d*. If the indexed sets $F(b,a)$ and $F(c,a)$ are already defined then the set $F(a,d)$ is defined as follows.

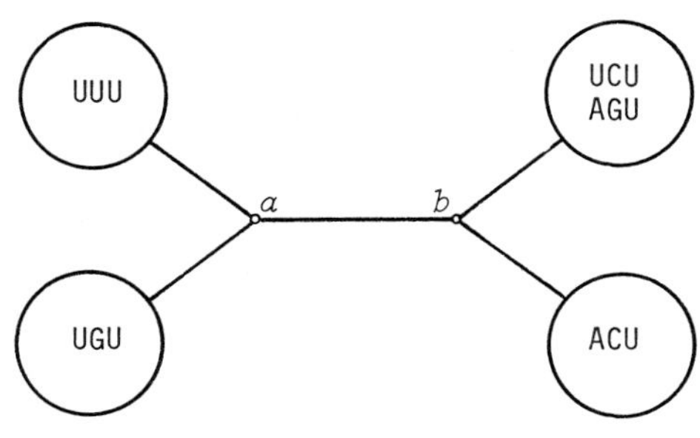

Figure 41

For each pair of codons $b_1b_2b_3 \varepsilon F(b,a)$ and $c_1c_2c_3 \varepsilon F(c,a)$ all possible codons $a_1a_2a_3$ are formed, where a_i is equal to either b_i or c_i $(i=1,2,3)$. The index of codon $a_1a_2a_3$ is set equal to the sum of the indices of codons $b_1b_2b_3$ and $c_1c_2c_3$ plus the distance between them. Then in the set thus obtained all indices are transformed by subtracting the value of the minimal index, with the subsequent removal of all codons which have a modified index greater than two. If there remain several instances of the same codon in a set, then only the one whose index is minimal is kept. In this way the set $F(a\ ,d)$ is completely determined.

It is known that in general $F(a,b) \neq F(b,a)$. In the example we are considering there are in all only two internal vertices. Clearly $F(a,b)$ = {UUU^0, UGU^0}, while $F(b,a)$ = {ACU^0, AGU^0, UCU^0}.

To form the set $F(a)$, the sets $F(b,a)$, $F(c,a)$ and $F(d,a)$ of the three vertices b, c, and d, adjacent to a, are used. For this we consider the set of all possible codons of the form $a_1a_2a_3$, where $a_i \in \{b_i, c_i, d_i\}$ for some $b_1b_2b_3 \in F(b,a)$, $c_1c_2c_3 \in F(c,a)$ and $d_1d_2d_3 \in F(d,a)$, with the codon $a_1a_2a_3$ being assigned an index equal to the sum of the indices of the initial codons plus the sum of the distances from them to $a_1a_2a_3$. The set $F(a)$ is formed from those codons with minimal indices in this set.

In our example $F(a)$ = {UUU, UGU, UCU}, $F(b)$ = {UCU, AGU}. Now the meaning of the codons in $F(a)$ becomes clear: a characteristic mapping can be realized only on them. Every other codon "accumulates" a larger distance from pendant vertices. At the same time, for any codon in $F(a)$ one can find a system of codons from the remaining sets $F(b)$, which realizes a characteristic mapping. One can conduct this process in such a way as to remember the "minimal" systems of codons during construction of $F(a,b)$ and $F(a)$. In particular, the further bounding of the excess leads to the following rule: A pair of codons from the sets $F(a)$ and $F(b)$ for neighboring vertices a and b cannot appear in the solution, if the distance between them exceeds the sum of the index of some codon from $F(a,b)$ and the distance between it and a codon taken from $F(b)$.

In our example consider the codon UUU$\in F(a)$. In $F(b,a)$ there are three indexed codons: ACU^0, AGU^0 and UCU^0. Calculating their distances from UUU (plus the indices, which are zero), we obtain 2, 2 and 1, respectively. Consequently together with UUU only UCU from $F(b)$ can be used, since the distance from AGU to UUU exceeds 1. Similarly, only UCU$\in F(b)$ can appear with UCU from $F(a)$, and for UGU$\in F(a)$ the formulated rule does not prohibit any codon from $F(b)$. Thus we have limited the set of possible variants of mappings to four (Fig. 42). Examining the lengths of these mappings shows that they are all equal to three, and consequently Figure 42 shows the complete set of characteristic mappings for the tree of Figure 40.

It is somewhat easier to reconstruct the probable ancestral structure for nucleotide sequences (keeping in mind the possibility of independent replacements at separate positions). A method of solution has been given by Fitch [63] and Hartigan [70a]. This method also consists of two stages.

In the first stage the number of candidates $F(x)$ in a given position is calculated for each vertex of the tree $x \in X$. The values $F(a)$ of a given position of the nucleotide sequence are given for all $a \in A$, where A is the set of pendant vertices. If, for a given $x \in X$, the values $F(y_i)$ for all the direct offspring $y_1, \ldots, y_k$ are already calculated, then $F(x)$ is set equal to the set of all those values (of nucleotides) encountered in the sequence $F(y_1), \ldots, F(y_k)$ with maximal frequency. In the case of a dendrogram $k=2$, so that either $F(x) = F(y_1) \cap F(y_2)$, if $F(y_1) \cap F(y_2) \neq \phi$, or $F(x) = F(y_1) \cup F(y_2)$ for $F(y_1) \cap F(y_2) = \phi$. Thus all the $F(x)$ for $x \in X$ are determined.

The next stage fixes the final form of the characteristic mapping. To do this the set $F(x)$ is modified, beginning at the root. For an arbitrary vertex x and its direct offspring y set $F'(x) = F(y)$ if $F(y) \subseteq F(x)$; otherwise retain in $F(x)$ any one of the elements of the sequence $F(y_1), \ldots, F(y_k), F(y)$, which occurs with maximal frequency, where $y_1, \ldots, y_k$ are the direct offspring of x.

It should be noted that this very algorithm is applicable also to sequences of polypeptides, if one keeps in mind the possibility of independent and equi-probable replacements of individual amino acids by others.

For more realistic situations, when the initial sequences have different lengths and due to deletions and insertions the corresponding positions have not been established beforehand, the problem of reconstructing ancestral structure has been examined in [90,90a].

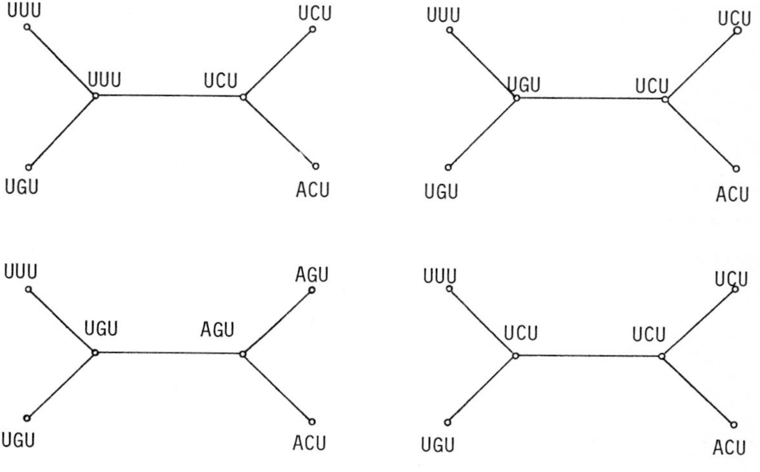

Figure 42

5. Calculating the internal structure of sequences during tree construction. The procedure we have described for constructing phylogenetic trees consists of sequential stages: calculation of the matrix of distances between the available genetic texts, formation of the tree topology, estimation of edge lengths, and the reconstruction of ancestral sequences, i.e. sequences corresponding to the internal vertices of the tree. However, considering that the original information available concerns sequences of amino acids or nucleotides, it is natural to try to concommitantly reconstruct both the dendrogram topology and the internal structures of ancestral sequences.

As a measure of the quality of the reconstructed evolution from a set of genetic texts one can use the total number of replacements necessary to generate these texts from a single root sequence. Minimizing this quantity we obtain an evolutionary tree most simply, i.e. by the most parsimonious method which explains the contemporary picture. The corresponding problem is called the problem of maximum parsimony. The principle of obtaining maximal parsimony, which we followed in the preceding section, has long been used in the investigation of the evolution of macromolecules [85,64a,6a].

The solution of the problem of constructing the most parsimonious tree has many interesting properties. We first note those contained in the very formulation of the problem, taking into account that chains of the tree obtained must indicate not only the fact, but the method of transition from one sequence to another, with edge lengths expressing the real differences between corresponding sequences. Consequently, in this problem edges with negative lengths are not allowed. On the other hand, some edges may connect identical sequences, thus having zero length. This possibility of zero-length edges permits the transformation of any tree into a binary tree, i.e. a dendrogram, by the addition of such edges. The optimal tree, then, may be constructed from the beginning as a dendrogram.

We note that for arbitrary contemporary species (i.e. pendant vertices of the tree) i and j the total number of replacements in the chain $[i,j]$ joining them may not be less than the distance d_{ij} between them, since they must be obtained in the tree from a common ancestral sequence by corresponding replacements, and must necessarily accumulate a distance d_{ij} in the process. In these circumstances, then, we always have

$$d_{ij} \leq d(i,j).$$

If in the constructed tree $d(i,j) - d_{ij} > 0$ then in the formation of the sequences i and j there must have appeared not only divergent, but

but also convergent or parallel processes in the evolutionary model represented by the tree. It is these processes that produce an increase in the total number of replacements in comparison with the value d_{ij}, which must have been obtained from only the influence of divergent processes.

We illustrate several, more subtle properties of the task of constructing the most parsimonious tree [6a,64a] in the following pre-arranged example of four nucleotide sequences:

	1	2	3	4	5	6	7	8
1	A	A	A	A	A	G	G	G
2	A	G	G	A	G	A	G	G
3	G	A	G	G	G	A	G	G
4	G	G	A	G	U	U	U	G

The matrix of distances which estimate the number of position-by-position disagreements between the given sequences is:

	1	2	3	4
1	0			
2	4	0		
3	5	3	0	
4	6	6	5	0

At first glance it seems that the sequences 2 and 3 must have a direct common ancestor, as they are the nearest to one another. However, this is not actually so: Their nearness is connected with redundant information, with the presence of those positions in the sequences which are non-essential, i.e. not informative for the construction of the evolutionary tree.

Notice first of all position 8, in which all the sequences contain the nucleotide G: Aside from dependence on tree topology no replacements of this nucleotide are necessary. In the optimal tree it is the same in all sequences, contemporary and ancestral. Clearly, in general, positions in which all the sequences being considered have the same symbol provide no information for construction of the tree.

Positions 5, 6 and 7 present another example of non-informativeness. They are distinguished by the property that in each of them only one symbol appears more than once: in positions 5 and 7 it is G, and in position 6 it is A. For such positions there must be no fewer replacements in the evolutionary tree than there are symbols which appear only once. But these replacements, independently of the topology of the optimal tree, are the most economically organized as follows: In all internal vertices of a given position that element must appear which occurred more than once in the corresponding sequence, unique symbols at pendant vertices being obtained from it by single replacements. Thus these positions provide no information for the formation of the tree.

It is clear that for more than four sequences at least one of the nucleotides will not be unique. With amino acids uniqueness of the symbols in a position is possible only for twenty or fewer sequences. Such a position, containing only uniquely occurring symbols also fails to have any influence on the topology of the tree, since any one of the symbols may be placed at the internal vertices.

The only positions, then, that are helpful in the construction of the optimal tree are those in which two or more symbols occur with a frequency greater than one [6a,64a]. The first four positions in our example are of this type. Removing the four non-informative positions, we have following distance matrix:

	1	2	3	4
1	0			
2	2	0		
3	3	3	0	
4	3	3	2	0

In agreement with this matrix, sequences 1 and 2 have a direct common ancestor, as well as sequences 3 and 4. The optimal tree is shown in Fig. 42a. From the figure it can be seen that use of the distance

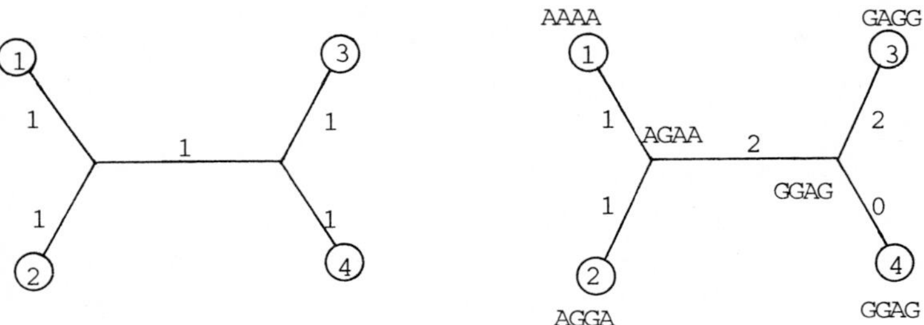

Figure 42a

matrix alone indicates only 5 replacements are necessary, while calculation of symbol content of the sequences shows that fewer than 6 replacements will not suffice.

We see that information about the sequences eases the construction of the optimal tree, but at the same time imposes a lower bound on the number of possible replacements.

A. A. Zharkikh established several additional properties associated with the construction of the most parsimonious trees, which can be recognized from the initial information [6a]. We will formulate several of these.

Identical sequences in an optimal tree have a common direct ancestor with exactly the same sequences attached to it, since this gives the minimal number (zero) of replacements. This property allows the deletion of identical rows, which may lead to an increase in the number of positions having only unique elements, i.e. in the number of non-informative positions. Repeated deletion of identical rows and of non-informative columns in individual cases may significantly reduce the size of the matrix.

The following conditions have a more subtle character, sufficient for joining or, on the other hand, separating individual vertices of a tree.

Let C be a subset of sequences. Then $n(C)$ will designate the number of positions, in each of which all the members of C are identical, but different from any sequence not in C. It can be shown that C forms a weighted subtree, if

$$n(C) > d_{ij}$$

for any $i,j \in C$. In particular, if $C = \{i,j\}$, then i and j have a direct common ancestor in the optimal tree, if they satisfy the inequality

$$d_{ij} > d_i + d_j,$$

where $d_i = \min_{k \neq i,j} d_{ik}$, $\quad d_j = \min_{k \neq i,j} d_{jk}$.

The proof of these properties is based on the fact that no edge of an optimal tree can be longer than the minimum of the distances between pendant vertices separated by that edge. Otherwise one could decrease the total length of the tree by throwing out that edge and joining, by a chain of shorter length, those pendant vertices (pairs) which attain the minimum d_{ij}.

The use of all these properties in the analysis of real data permits a significant reduction in the dimensionality and range of selection of possible variants [6a]. In the case of polypeptide sequences, non-unique optimal reconstructions of the tree and of ancestral proteins can be partially reduced by minimizing differences in the corresponding polynucleotide texts.

§2. The evolution of families of synonymous proteins

1. The dendrogram of the globins and its analysis. The methods described above have been used recently for the analysis of a whole series of protein families (See for example [58]). By *protein family* is meant a group of synonymous (iso-functional) polypeptides [31,98].

The *globin family* has been studied the most thoroughly. This family of proteins is responsible at the molecular level for the binding, transfer and release of molecules of oxygen and carbon dioxide. Up to the present more than 200 representatives of this family have been deciphered (from the globins of the mosquito to those of man). Moreover, in several cases (myoglobin of the sperm whale, hemoglobin of horse and man, globin of the worm and others) the spatial organization and chemical functions of these macromolecules is known in detail down to the atomic coordinates. In Section 2.1.2 we already dealt with the possibilities of describing the semantics of hemoglobin molecules.

A few words about this family. Within it there is notable a group of myoglobins, proteins active in muscle tissue. These are connected with the storage and consumption of energy in muscles, the functioning of which requires internal energy reserves. It is the myoglobin molecules that serve as this distinctive oxygen depot. Unlike the hemoglobins, myoglobin chains do not form quaternary structures, but function in the form of separate monomers.

Aside from myoglobin cistrons, in higher mammals there are several cistrons coding for various forms of hemoglobin chains, designated, for example, in man as $\alpha,\beta,\gamma,\delta$ and ε. These chains function in erythrocytes of the blood, in different periods of an individual's growth (*ontogenesis*). The hemoglobin of an adult person consists of 2α- and 2β-chains, as was already noted in Section 2.1.2. However in certain periods of ontogenesis in place of the β-chains in the tetrameric molecule, chains of other hemoglobin cistrons are used: ε-chains are encountered only in the earliest stages of embryonic growth, γ-chains are included in hemoglobin used during most of the interuterine growth, but soon after birth it disappears completely and with this change the embryonic tetramer $2\alpha 2\gamma$ is replaced by adult hemoglobin $2\alpha 2\beta$. Finally, in some postembryonic molecules the δ-chain replaces the β-chain; though the concentration of such proteins is normally very small. Thus one commonality of observed hemoglobin fractions is the presence of products of the α-cistrons. The products of the remaining hemoglobin genes (β-, γ-, and δ-chains) possess remarkable similarity to the α-chain as well

---

†The primary structure of ε-chains has not been completely deciphered, and therefore we will omit them from further discussion.

as among themselves. This gives reason to believe that all known contemporary globin chains arise from a common ancestor.

In recent years impressive success has been obtained in deciphering the structure and organization of the genes which code for globin sequences (basically in mammals: man, rabbits, mice and others), see for example the survey [82a]. It turns out that at the chromosome level the globin genes are concentrated in compact clusters, in which their order of appearance reflects the order of the functioning in the course of ontogenesis. The cluster of β-like cistrons for synthesis of hemoglobins is shown in Fig. 43' (the scale is given in terms of bases and kilobases). First note that the cluster consists of the translatable genes $\varepsilon, G_\gamma, A_\gamma, \delta, \beta$ and the nontranslatable genes $\psi\beta 1, \psi\beta 2$, called pseudogenes. Remarkably, these very pseudogenes divide the fragment into "embryonic" and "postembryonic" sections, including genes active at the embryonic stage, and genes functional only after birth of the organism. In passing, we note that pseudogenes differ from "normal" genes in structure: they have no introns, while usually intron portions of genes constitute a very significant fraction of the whole. For an example see Fig. 43',a, where the intron segments are unshaded. This serves as still another indirect argument in support of the conception of the regulatory role of introns in the reading-out processes.

A special group is constituted by the "respiratory" molecules in insects, lampreys and fish. Limited understanding of the respiratory

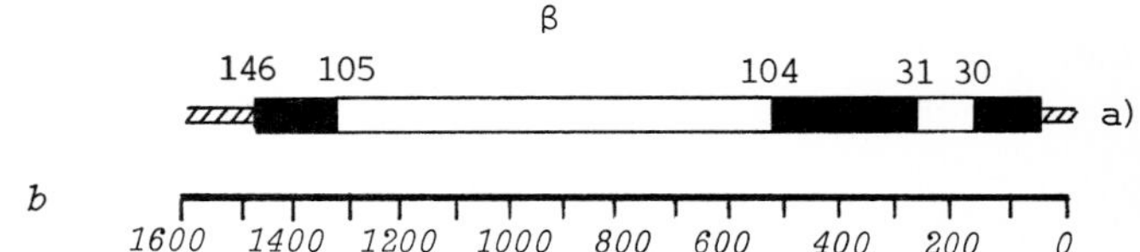

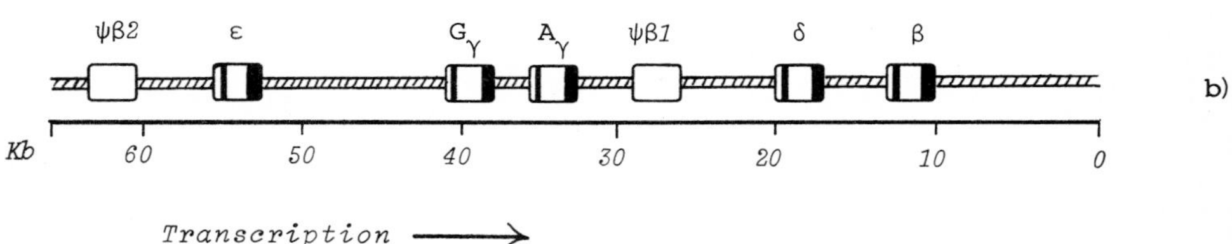

Figure 43'. **Structure of human β-hemoglobin gene (a) and linkage arrangement of all human β-like genes in a chromosome.** The black and white boxes represent the coding (exon) and non-coding (intron) sequences, respectively. The length of introns is equal to approximately 125-150 and 800-900 base pairs (b), located between codons 30 and 31 and 104 and 105, respectively. The scheme b) shows the positions of the embryonic (ε), fetal (G_γ, A_γ), and adult (δ, β) β-like globin genes and two β-like pseudogenes ($\psi\beta 1, \psi\beta 2$). The length of the entire β-like globin cluster is of 50-60 thousands of base pairs (i.e., 50-60 Kb in the figure).

function of these animals does not permit comparison of the molecules with those groups enumerated above, and so we will call them simply globins.

The collection of primary structures of the globin family which have been studied is sufficiently rich that one can suppose the corresponding dendrogram to reflect all the fundamental stages in the history of this protein.

In Fig. 43 is shown a dendrogram constructed for the matrix of minimal mutational distances of 68 different globin sequences using the unification algorithm [69]. The parts of the tree corresponding to the evolution of α-chains and β-like chains were locally reconstructed using the Moore coefficient [69]. For our purposes the numeric values of edge lengths were non-essential, and they are not included. However, the figure does reflect the relative lengths. Negative lengths were obtained for some edges, and these are shown in the figure with "reverse" directions: upward from the point of branching. Negative edges are few, and their lengths are not great, as one should expect from our previous understanding of the reasons for their appearance.

We first note that the dendrogram of Fig. 43 is in close agreement with classical representations of evolutionary nearness of the species under consideration, which are based on the analysis of external, morphological traits.

We now turn to a more detailed analysis of the branching structure of this tree, in accordance with [30,31]. The symbols d_1, d_2, d_3 and d_4 designate the major stages in the evolution of the globin family.

At d_1 the family is divided into two groups: a group of insect globins (1,2) and a group of vertebrate globins (3-68). At d_2 the vertebrates divide into the myoglobins (3,4) and lamprey globins (5), and the group of hemoglobins proper (6-68). At d_3 the latter group is divided into the α-chains and the β-like chains. The β-like chains are divided at d_4 into the group of γ-chains and group of β- and δ-chains of mammals. These branch points can be naturally interpreted as stages, leading to the appearance of new genes coding for new variants of hemoglobin chains. This is not true of d_1, since we still are not sure it represents the appearance of globin genes. The remaining branch points of the tree correspond to divergences of the same type of chain, but in different species.

To clarify the foregoing we note that the basis of cistron divergence is the independent accumulation of mutational replacements during

---

†The alert reader will easily notice the non-uniformity of representation of the various species in the tree. This is due to the fact that the α- and β-chains of mammalian hemoglobin are the most heavily studied.

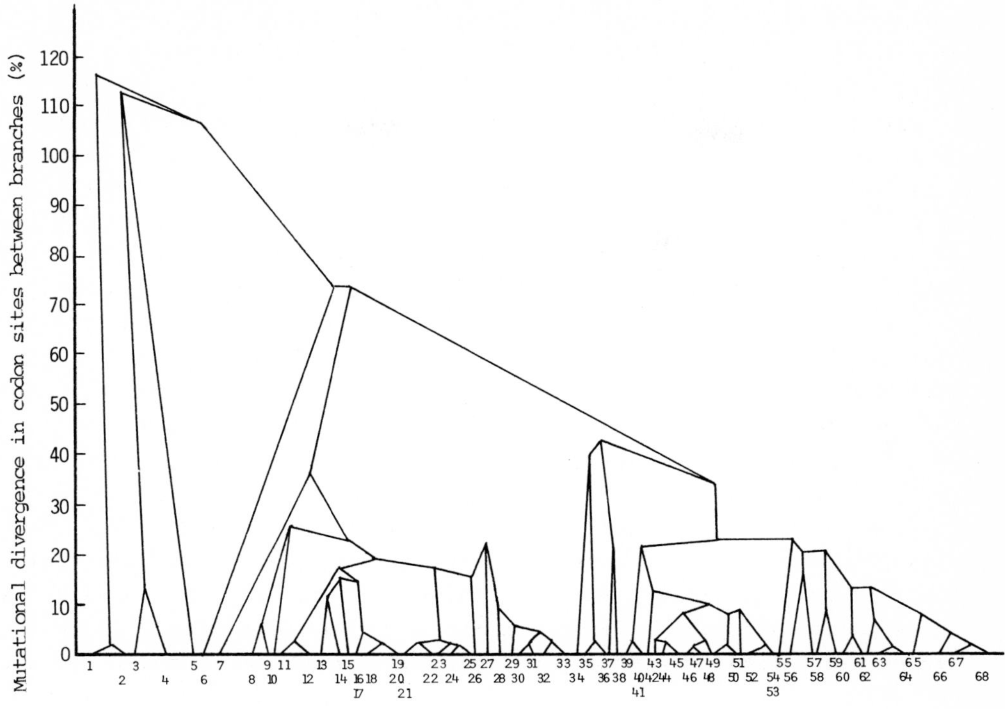

Figure 43. The dichotomous globin tree [69]

1 - globin *Chironomus thummi*; 2 - allelic variant, globin *Chironomus thummi*; 3 - myoglobin sperm whale; 4 - myoglobin horse; 5 - globin lamprey; 6 - globin carp; 7 - α chicken; 8 - α rabbit; 9 - allelic variant α rabbit; 10 - α tree shrew; 11 - α mouse NB; 12 - α mouse C-57 Bl; 13 - α sifaka; 14 - α lemur; 15 - α *Galago crassicandatus*; 16 - α *Macaca mulatta*; 17 - α *Macaca fuscata*; 18 - α gorilla; 19 - α chimpanzee; 20 - α human; 21 - α donkey; 22 - α horse slow; 23 - α horse fast; 24 - α horse fast, allelic variant; 25 - allelic variant α horse slow; 26 - α pig; 27 - α llama; 28 - α bovine; 29 - α goat; 30 - allelic variant α goat; 31 - α goat, fraction A; 32 - α sheep, fraction D; 33 - α sheep, fraction A; 34 - γ human; 35 - β kangaroo; 36 - allelic variant β kangaroo; 37 - β sifaka; 38 - β lemur; 39 - β mouse AKR; 40 - β mouse Sec; 41 - β mouse C-57 Bl; 42 - β rabbit; 43 - β squirrel monkey; 44 - β tamarin; 45 - β *Ateles geoffroyi*; 46 - δ squirrel monkey; 47 - δ tamarin; 48 - δ spider monkey; 49 - β *Macaca fuscata*; 50 - β *Macaca mulatta*; 51 - δ human; 52 - β gorilla; 53 - β chimpanzee; 54 - β human; 55 - β horse, slow fraction; 56 - β pig; 57 - β llama; 58 - β bovine foetal; 59 - β sheep foetal; 60 - β bovine, fraction A; 61 - β bovine, fraction B; 62 - β sheep barbary, fraction C; 63 - β goat, fraction C; 64 - β sheep, fraction C; 65 - β sheep, fraction B; 66 - β sheep, fraction A; 67 - β goat, fraction A; 68 - β goat, fraction A, allelic variant.

evolution. We first examine the so-called *inter-species divergence*, in which hereditary differences are accumulated in descendants of the same gene, separated by an isolating barrier. In this situation the ancestral population is clearly divided into a number of subpopulations, isolated from one another in a genetic sense. Since the subpopulations do not mix there is no exchange of genetic material among them. The isofunctional cistrons of each population independently accumulate various mutations. As a result of this process in contemporary species we find amino acid differences among proteins which perform identical functions. It is known that the number of subpopulations is not necessarily two, so that the notion of a dendrogram (corresponding to dichotomous divergence) is not adequate for the process of interspecies divergence.

Divergence associated with the appearance of new genes (points d_2, d_3 and d_4 in Fig. 43) is called intra-species divergence. A possible mechanism for intraspecies divergence in connection with hemoglobin evolution was first described clearly in the work of V. Ingram [74]. From his study of globin chain similarities in humans he proposed that the appearance of new globin cistrons in organisms is explained by the duplication of old copies obtained by successive, independent evolutionary processes [74,30,31]. In this case genetic evolution results from the relative inability of the duplicated genes to recombine, since they are non-homologous.

If the evolutionary process were limited only by interspecies divergence, the appearance of new functions (while retaining the old), which leads to more complex organization, i.e. to *progressive* evolution [27], would be impossible: New functions must be represented in the genetic memory by new genes (a determining, though lesser, role may be played by other mechanisms as well [98]). Duplicate genes may accumulate different mutations, permitting a gradual transition to new functions.

Though gene duplication is an extremely rare phenomenon, rarer still is the appearance of more than two copies of a gene. This is the reason that intra-species divergence is adequate for the description of the dendrogram.

The points d_2, d_3 and d_4 in Fig. 43 correspond to major duplications of pre-cistronic globins. First (d_1) the globins of vertebrates and nonvertebrates diverged (evidently without duplication), then through duplication the myoglobins were separated out (d_2), then the pre-cistron for the hemoglobins diverged, with subsequent divergence of α- and β-chains (d_3); d_4 corresponds to a duplication for the ancestral β-cistron, as a result of which the γ-chains of mammals were isolated.

The question arises, what is the evolutionary meaning of these divergences? Do these hypothetical duplications represent real ages in the history of living forms?

To answer these questions our dendrogram, in which time is measured indirectly by the number of fixed amino-acid replacements, must be connected with real geologic time. An initial premise for such a connection is the postulate that amino acid differences in comparable contemporary proteins are greater, the earlier they diverged. The most simple thing would be to assume that the number of differences is proportional to the time of the independent evolutionary processes of these proteins. Then to estimate the times for all the branch points of the tree it would only be necessary to know the rate of fixation of any single replacement. To determine this rate one can use reliable independent paleontological data concerning the divergence times of some species, i.e. the times since their common ancestor.

However such a calculation does not take into account the important fact that replacements in a single position of a chain can occur (and have occurred) repeatedly. To include this circumstance in our calculations consider the following reasoning [99,30,31]. Let λ be the average number of replacements in a given position per unit of time. Since the appearance of a mutation is an event of small probability, it is natural to assume that the occurrence of repeated replacements in a given position is described by a Poisson distribution, so that the probability of n replacements in time T is given by

$$p_n(T) = \frac{(\lambda T)^n e^{-\lambda T}}{n!}.$$

The probability of no replacements in time T is then

$$p_0(T) = e^{-\lambda T}.$$

The number of coinciding positions in two sufficiently long polypeptide chains which have diverged in time T gives an estimate of this quantity. We designate the fraction of differing positions by Q, so that

$$Q \simeq 1 - p_0(T) = 1 - e^{-\lambda T}.$$

For $\lambda T << 1$ the approximating equality

$$Q \simeq \lambda T$$

holds. This formula permits the estimation of λ by means of those chains which have separated comparatively recently (in geologic terms). This is the method used by E. Zuckerkandl and L. Pauling [99]. They found that between the α- and β-chains of hemoglobin in swine, horses, oxen and rabbits, on the one hand, and the corresponding chains in humans, on the other, there are an average of 22 amino acid differences. From paleontologic data it is known that the common ancestor of these mammals lived approximately 80 million years ago, so that $Q = \frac{22}{145} = 0.15$ (145 is the average length of the chains being compared), $T = 8 \cdot 10^7$ years. From this $\lambda \simeq 2 \cdot 10^{-9}$ mutations per year.

Substituting this value of λ in the formula for $p_0(T)$ gives the *Pauling scale* for the translation of mutational differences into units of real time (Fig. 44). The function $Q(T)$ is graphed in this figure, and several points of interest are marked on it [30,31]. The point d_2 ($Q = 0.75$) corresponds to 650 million years, the end of the Pre-Cambrian, when invertebrate sea aminals appeared. Consequently in this period specification of the energetic process in muscles had already occurred, as well as in other tissues (tissue respiration as distinguished from pulmonate). Point d_3 (the average fraction of differences between α- and β-like chains, $Q = 0.52$) corresponds to the Devonian period, about 380 million years ago, when vertebrates moved onto dry land. It was then, apparently, that the cleavage from the former respiratory mechanism occurred -- the transition to oxygen respiration from air by means of lungs. The point d_4 ($Q = 0.26$) corresponds to T = 150 million years, the Jurassic period, and the appearance of marsupials. The separation of γ-chains is associated with the singling-out of the embryonic stage in the growth of mammals. The substitution of γ-chains for β-chains in embryonic hemoglobin is evidently connected with the fact that oxygen enters the embryo through the placenta from the mother's blood, and the function of hemoglobin $2\alpha 2\gamma$ is to take oxygen away from the $2\alpha 2\beta$ hemoglobin of the mother.

Thus the evolution of globin structures as a whole, in accord with the tree of Fig. 43, correlates well with phenotypic modifications and restructurings of the respiratory function. Moreover, by the method of constructing such trees one can investigate the real, molecular state of affairs in evolution.

Having satisfied ourselves that the construction of the dendrogram as a whole is adequate, we now turn to a more detailed analysis of the globins in various sections of the dendrogram.

It has been found that the rate of fixation of mutations, calculated for various sections of the protein phylogenetic tree, is nearly constant, even though these sections are associated with strongly dis-

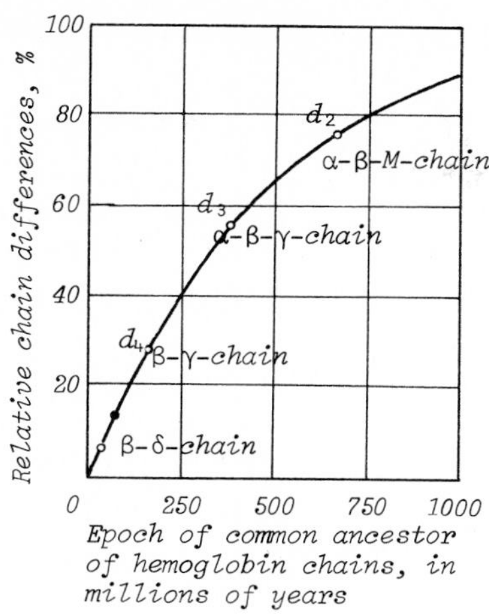

Figure 44

tinguished environmental periods and intense morphological evolution. For example the average mutational distance from carp hemoglobin to the α- and β-chains of humans is nearly the same as the distance between these α- and β-chains [76,77,32], so that the fixation rates of mutations in the lines leading to carp hemoglobin and the α- and β-chains of humans from their common ancestor are practically the same. However, as is well known, contemporary carp differs little from this ancient ancestor morphologically (its dwelling conditions have changed little in 400 million years), while in those lines leading to man there have been violent bursts in the generation of forms under the impetus of major variations and sharp changes in environmental conditions. This leads one to the conclusion that molecular changes in globin genes are weakly connected with the progressive evolutionary complexity of organization in living systems: The average rate of fixation of replacements in proteins does not depend on other conditions, such as ecological ones.

This conclusion appears paradoxical, since it contradicts the idea of genetic change as the basis of all evolutionary transformations. The paradox was destroyed by several scientists who hypothesized non-Darwinian gene evolution, according to which, the overwhelming majority of mutations fixed in a population during the course of evolution are non-adaptive, i.e. neutral (they make little change in the fitness of the organism for its surroundings), from which, in the framework of the genetic theory of populations, it is concluded that the average rate of fixation of mutations is constant [76,77,32].

2. Analyzing the evolution of globin sequences from their internal structure. The constancy of the average fixation rate of protein replacements, discovered in the analysis of phylogenetic trees constructed on the basis of distance matrices, may be considered an artifact. Within the approach itself lie tendencies toward smoothing of differences in the rate of the evolutionary divergence of species: first, from the transition to summed characteristics in measuring the nearness of sequences in the distance matrix, and second, from the use of the agglomerative unification procedure, which is tightly bound up with

properties of the ultrametric associated with the uniform flow of the evolutionary process.

To clear up this important question in the theory of evolution the alternate approach of Fitch and Zharkikh can be used. In this approach the reconstruction of a phylogenetic tree is based on the direct position-by-position comparison of protein primary structures (see Section 3.1.5). We will describe the results obtained in the paper [34a] by the use of this approach.

Figure 45 shows the phylogenetic tree for 159 globin sequences: 49 α-like, 72 β-like, 36 myoglobin chains and 2 lamprey globins. This collection is much more complete than the one described in the preceding section, but unfortunately it does not encompass the other one completely (missing in particular are the globins of invertebrate animals).

We first compare the structures of the trees in Figs. 43 and 45. Against the background of general agreement a number of differences appear, perhaps the most notable of which is the following: In the new tree the line for the α-like chain of carp is singled out later than the duplication of the ancestor cistron for α- and β-like chains, while in the tree of Fig. 43 this event precedes the duplication point. The variant of Fig. 45 appears better grounded, since apart from all the others the α-like carp chain has a deletion in common with all the other α-like chains of animals, at the time when there is no such deletion in the β-cistrons. The importance of this consideration flows from the fact that the fixation of such large-scale mutations as deletions is a very infrequent event.

A second source of differences stems from the fact that unique replacements in the latter approach are located on pendant edges of the tree, which is also more valid. Thus the two variant α-chains of goat hemoglobin (Fig. 45) differ by 5 unique replacements, all of them located on one branch of the tree, which originates at the direct common ancestor (which therefore coincides with the second of these sequences). Its nearest neighbors, the α-chains of sheep, differ by two or three replacements in all, and, by the logic of algorithms based only on the distance matrix, must be wedged in between the indicated chains of goat hemoglobin, which is patent nonsense. By similar reasoning one can explain several curiosities in the taxonomic arrangement of individual species in trees constructed for other protein families on the basis only of distance matrices. Thus, in the cytochrome tree from [65] the tortoise is closer to birds than to snakes. But if unique replacements are excluded, the sequences of all these species fall together. Most likely, these species diverged from a common ancestor in a non-dichotomous group, but with different velocities: The more rapid accumulation

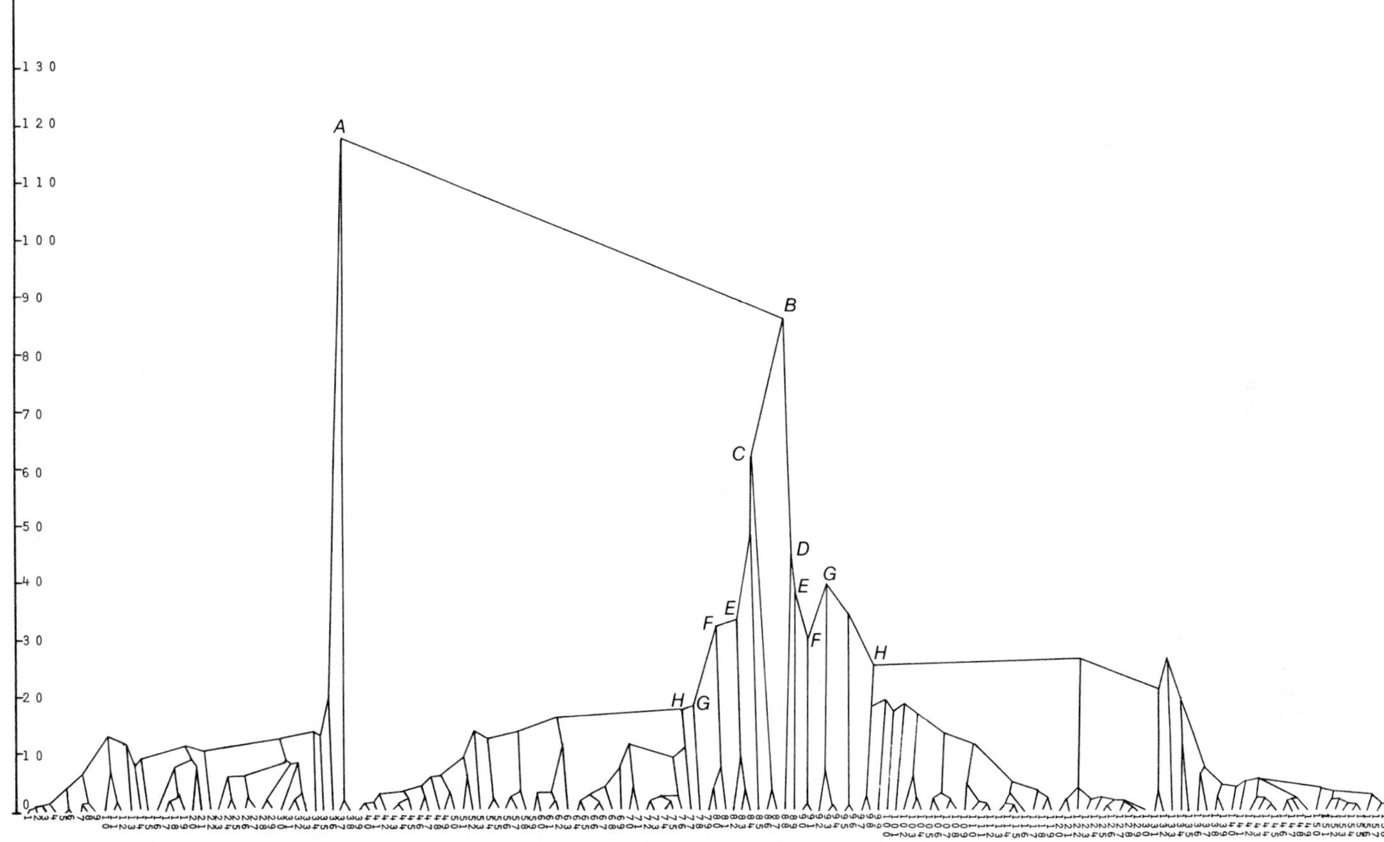

Figure 45. The phylogenetic tree of the globin superfamily constructed by the method of A. A. Zharkikh.

Figure 45 (legend): Designations: A,B,C... - some key points of divergence. The values for rates of substitution were calculated by adding an independent paleontological time scale to the following points: A - the root of the tree (ancestor of vertebrates), C - teleost ancestor, D - tetrapod ancestor (first amphibia), E - amniote ancestor (the point of divergence of birds and mammals), F - mammalian ancestor, H - common ancestor of contemporary mammalian orders; B corresponds to duplication of pre(α- β) cistron, i.e. to divergence of α- and β-hemoglobin chains. Other points mark some important divergences in primate phyletic line.

I. Myoglobins. 1- Human; 2- Chimpanzee, *Pan troglodytes*; 3- Gorilla, *Gorilla gorilla*; 4- Gibbon, *Hylobates agilis*; 5- Olive baboon, *Papio anubis*; 6- Irus macaque, *Macaca fascicularis*; 7- Woolly monkey, *Lagotrix lagothricha*; 8- Squirrel monkey, *Saimiri sciureus* ; 9- Common marmoset, *Callithrix jacchus*; 10- Thick tailed galago, *Galago crassicaudatus*; 11- Potto, *Perodicticus potto*; 12- Slow loris, *Nycticebus coucang*; 13- Sportive lemur, *Lepilemur mustelinus*; 14- Common treeshrew, *Tupaia glis*; 15- Hedgehog, *Erinaceus europeaus*; 16- Badger, *Meles meles*; 17- Fox, *Vu;pes vulpes*; 18- Domestic dog, *Canis familaris*; 19- Wild dog, *Lycaon pictus*; 20- California sea lion, *Zalophus californianus*; 21- Harbor seal, *Phoca vitulina*; 22- Rabbit; 23- Sperm whale, *Physeter catodon*; 24- Common dolphin, *Delphinus delphis*; 25- Black Sea dolphin, *Tursiops truncatus*; 26- Amazon River dolphin, *Inia geoffrensis*; 27- Common porpoise, *Phocoena phocoena*; 28- Horse; 29- Zebra; 30- Pig; 31- Sheep; 32- Red deer, *Cervus elaphus*; 33- Bovine; 34- Red kangaroo, *Megaleia rufa*; 35- Opossum, *Didelphis marsupialis*; 36- Chicken, *Gallus gallus*; 37- Lamprey globin, *Lampetra fluviatilis*; 38- Sea lamprey globin, *Petromyzon marinus*.

II. Hemoglobin, alpha-chain. 39- Human; 40- Chimpanzee, *Pan troglodytes*; 41- Gorilla, *Gorilla gorilla*; 42- Hanuman langur, *Presbytis entellus*; 43- Rhesus monkey, *Macaca mulata*; 44- Japanese monkey, *Macaca fuscata fuscata*; 45- Savannah monkey, *Cercopithecus aephips*; 46- Capuchin monkey, *Cebus apella*; 47- Spider monkey, *Ateles geoffroyi*; 48- Tarsius, *Tarsius bancanus*; 49- Bush baby, *Galago crassicaudatus*; 50- Slow loris, *Nycticebus coucang*; 51- Sifaka, *Propithecus verreauxi*; 52- Brown lemur, *Lemur fulvus*; 53- Tree shrew, *Tupaia glis*; 54- Dog, I; 55- Dog, II; 56- Rabbit, I; 57- Rabbit, II; 58- Rabbit, III; 59- Mouse, I; 60- Mouse, II; 61- Mouse, III; 62- Mouse, IV; 63- Rat, *Rattus norvegicus*; 64- Goat, A; 65- Barbary sheep, *Ammotragus lervia*; 66- Sheep, A; 67- Sheep, D; 68- Goat, B; 69- Bovine; 70- Pig, I; 71- Pig, II; 72- Donkey; 73- Horse, I; 74- Horse, II; 75- Horse, III; 76- Horse, IV; 77- Llama, *Lama peruanà*; 78- Gray kangaroo, *Macropus giganteus*; 79- Echidna, I, *Tachyglossus aculeatus aculeatus*; 80- Echidna, II; 81- Echidna, III; 82- Goose, *Anser anser*; 83- Chicken, I; 84- Chicken, II; 85- Viper, *Vipera aspis*; 86- Carp, *Cyprinus carpio*; 87- Catostomus clarkii.

III. Hemoglobin, beta-chain. 88- Frog, *Rana esculenta*; 89- Chicken; 90- Echidna, *Tachyglossum aculeatum*; 91 Echidna, II; 92- Potoroo, *Potprous Tridactylus*; 93- Red kangaroo, *Megaleia rufa*; 94- Gray kangaroo, *Macropus giganteus*; 95- Human gamma-chain, I; 96- Human gamma-chain, II; 97- Rabbit, I; 98- Rabbit, II; 99- Horse; 100- Llama, *Lama peruana*; 101- Pig; 102- Bovine, F; 103- Goat, F; 104- Sheep, F; 105- Bovine, D; 106- Bovine, C; 107- Bovine, A; 108- Bovine, B; 109- Barbary sheep, C(NA); 110- Goat, C; 111- Barbary sheep, C; 112- Sheep, C; 113- Goat, E; 114- Goat, D; 115- Goat, A; 116- Sheep, B; 117- Barbary sheep, B; 118- Sheep, A (Soay breed); 119- Sheep, A (Clan breed); 12- Mouse, I; 121- Mouse, II; 122- Mouse, III; 123- Mouse, IV; 124- Mouse, V; 125- Mouse, VI; 126- Mouse, VII; 127- Mouse, VIII; 128- Mouse, IX; 129- Mouse, X; 130- Mouse, XI; 131-

Badger, *Meles meles*; 132- Dog; 133- Treeshrew, *Tupaia glis*; 134- Brown lemur, *Lemur fulvus*; 135- Sifaca, *Propithecus verreauxi*; 136- Bush baby, *Galago crassicaudatus*; 137- Slow loris, *Nyctycebus coucang*; 138- Ramarin, *Saguinus mystax*, *S.nigricollis*; 139- Capuchin monkey, *Cebus apella*; 140- Night monkey, *Aotus trivirgatus*; 141- Squirrel monkey, *Saimiri sciureus*; 142- Spider monkey, delta-chain, *Ateles geoffroyi*; 143- Night monkey, delta-chain, *Aotus Trivirgatus*; 145- Squirrel monkey, delta-chain, *Saimiri sciureus*; 146- Gibbon, delta-chain, *Hylobates lar*; 147- Gorilla, delta-chain, *Gorilla gorilla*; 148- Chimpanzee, delta-chain, *Pan troglodytes*; 149- Human, delta-chain; 150- Spider monkey, *Ateles geoffroyi*; 151- Hanuman langur, *Presbytis entellus*; 152- Savannah monkey, *Cercopithecus aethiops*; 153- Rhesus monkey, *Macaca mulata*; 154- Japanese monkey, *Macaca fuscata fuscata*; 155- Irus monkey, *Macaca irus*; 156- Gibbon, *Hylobates lar*; 157- Gorilla, *Gorilla gorilla*; 158- Chimpanzee, *Pan troglodytes*; 159 Human.

of unique mutations in snakes led to the mistake in the taxonomy.

Since in Fig. 45 ancestral sequences were constructed simultaneously, so that edge lengths accurately reflect the number of replacements among them, one can try to estimate the rate of the evolutionary process at individual stages of evolution, connected with those points of divergence for which more or less accurate paleontologic data are known. These points are: A - the root of the tree, the common ancestor of vertebrates (about 500 million years ago), C - the common ancestor of fish and land vertebrates (about 400 million years back), D - the common ancestor of tetrapods (the first amphibian, existing approximately 340 million years ago), E - the common ancestor of amniotes, the point of divergence of birds and mammals (about 300 million years ago), F - the common ancestor of mammals (about 150 million years ago), G - the common ancestor of placental mammals (about 120 million years ago) and H - the common ancestor of the principle contemporary order of mammals (about 75 million years back).

On the basis of these data in the paper [34a] the average rates of evolution of the globins were estimated in sections of the tree situated between neighboring pairs of these distinguished points, using the Pauling scale. These rates were calculated not only for the molecules as wholes, but also for the major substructures: the semantic and non-semantic sections, with additional estimates being made for the individual functional centers: the heme-specific center, the centers of contact and the center of binding of 2,3-diphosphoglycerate.

The calculated speeds are far from uniform, some values being as much as 20 times larger than others. The greatest speed was found in the period between 500 and 300 million years ago, when the quaternary structure of hemoglobin was being formed: the contact centers and allosteric centers. The rates of fixation of replacements in the functional centers were found to be several times larger than those in the non-semantic sections. The rates of change were particularly high in the centers which ensure cooperative effects in the functioning of the tetramer. From paleontologic data, it is this period that saw the emergence of land vertebrates and the transition to oxygen respiration from air.

In the interval from 300 to 150 million years ago the evolutionary process of hemoglobins slowed sharply, with the most conservative sections being the centers that ensure cooperative properties (the heme-specific center, the contact α_1-β_2 and the center of binding of 2,3-diphosphoglycerate). Not one mutation became fixed in these centers throughout this period. Finally, in the last 150 million years the fixation rates in various branches of the tree fluctuated less strongly, averaging considerably lower than in the period of emergence of land vertebrates. On the average β-chains evolved more rapidly than α-chains in this later period. A similar result was obtained in [79].

We will consider these results in the light of evolutionary and genetic ideas.

First of all our attention is called to the concurrence of three different processes: the increased rate of accumulation of replacements in the earliest stage of the tree shown in Fig. 45, the formation of quaternary structure in hemoglobin, and the coping with the fundamentally new ecological conditions associated with land dwelling -- the transition to atmospheric respiration of oxygen. The point B marks the duplication of the pre-cistron of α- and β-chains, with their subsequent mutational divergence (the point d_3 of Fig. 43 corresponds to it). This point corresponds to the formation of developed quaternary structure in hemoglobins, since all the earlier divergent globins normally operated in the form of protomers, on the level of tertiary structure. The transformation of globins to quaternary structure at this stage is completely "logical", since it is such structures that bring about cooperative functional effects necessary for living in a high-oxygen environment [32]. It should be remembered that the multimeric structure of hemoglobin provides not only for the cooperative binding of oxygen molecules, but also for their release, under imposed conditions, more easily than with protomers. This feature "rescues" animals from the effects of large changes in partial oxygen pressure [49].

Strictly speaking, the appearance of quaternary structure in globins should be dated somewhat earlier than the point d_3, when identical globin chains were able to aggregate, forming multimers with weak cooperative functional effects. However, the corresponding co-adaptive aggregating variants of globin chains lost one another as a result of meiosis [43]. Duplication of the pre-cistron of the α- and β-chains secured this initial co-adaptivity, which up to then had been incomplete. The duplicated genes were then able to evolve independently. And in fact for more than 300 million years the α- and β-chains of, for example, man have accumulated a significant number of amino-acid differences. But are these differences really independent? It is clear that the contacts α_1-β_2 and α_1-β_1 which determine, in essence, the quaternary hemoglobin structure, were under the particularly strict "supervision" of natural selection: If serious mutations arose in the centers of contact of the protomers then they became fixed as quickly as possible according to the principle that "a defect corrects a defect" (see Section 2.1.3), which was supported in the paper [36] by the analysis of the structures of contact centers at key branch points of the tree.

The higher rate of evolution in this earlier period can be understood if one considers that the rate of fixation of an arbitrary mutation in a population depends on the degree of fitness of its carriers for environmental conditions. The emergence of land vertebrates represented a sharp change in the ecological environment, and the protomeric globins, evolutionally fitted for a water invironment, lost their value in these fundamentally new circumstances. Among the spontaneously-occurring mutations at this time there must have arisen some fraction of alleles more suited to the new surroundings, i.e. there would be a sharp reduction in the time expected for the appearance of new variants, more preferred than the others. Thus the quickly evolving forms survive, their evolution having a clearly adaptive, "Darwinian" character. In a certain sense this evolution is progressively directed, since it is aimed at adapting to such a long-acting environmental factor as high partial pressure of oxygen. At the molecular level this is manifest in the improvement of quaternary structure of globins and the evolutionary adjustments in their cooperative functions.

The situation is quite different when the quaternary structure is already formed and is coping adequately with its functions. Under these conditions the chances of meeting more preferred variants arising from new mutations is lowered, and a slowing of hemoglobin evolution occurs. From this point the "optimized" quaternary structure of the globins is under the strict control of natural selection, which is now of a stabilizing nature. In this situation neutral or nearly neutral synonymous mutations may be fixed (by means of random genetic drift [31]), being manifest very little in protein structure and functions, and therefore not amenable to natural selection. Such evolution, generally speaking, has no expressed direction and can be characterized by quite large fractions of neutral replacements. For this reason it is called neutral evolution.

The relatively higher rate of evolution of β-like chains in comparison with α-like chains is easily explained by their different functional loads in ontogenesis: α-chains are involved in all hemoglobin fractions, i.e. they are active at all stages of ontogenesis, while the β-like chains $\beta, \gamma, \delta, \varepsilon$ interact only with α-chains, not among themselves. In general, the increased diversity of the β-like chains through the associated duplication of ancestral genes (Fig. 43',a) must be accompanied by a quickening of their evolution. But the differences they have accumulated as well as the resolving ability of our methods are not sufficient for a thorough-going analysis of this question. It can be conjectured, however, that beginning with the point of divergence of β- and δ-chains (the formation of the β and δ cistrons), replacements in the α cistron became more strictly evaluated by natural selection (the difference in the rate of entrainment of the positions of the β- and δ-chains grew precisely following this duplication [34a]).

The primary structures of α-, β-, γ- and δ-globins of mammals are sufficiently similar that one can assume homologous positions are involved in their centers of mutual recognition (contact) [30,31]. But then "serious" replacements in the contact centers of the α-chains could hardly become fixed, since for this to happen "simultaneous" co-adaptive mutations would have to occur in nearby positions of β-, γ- and δ-cistrons. The probability of such a coincidence of replacements in several cistrons at once is practically zero.

Thus this evolutionary feature of α-cistrons in mammalian hemoglobin is one more example of global restrictions on the semantic evolution of proteins.

Since Darwin's time a basic problem of the theory of evolution has been the mechanism for the formation of species. Are the changes in globin protein structures connected with the emergence and divergence of species on the morphological level? None of the foregoing bears directly on this question, since it is not connected with the formation of species, as such, but with the increasing complexity of organisms. The fundamental, tetrameric organization of hemoglobins with developed cooperative properties is characteristic, for example, for all contemporary species of mammals from the shrew to man. Can it be that, the "finest hour" of hemoglobin (the adaptive, quickened formation of quaternary structure) having passed, during the next two to three hundred million years it actually evolved under a neutral regimen? One might answer affirmatively, in view of the relatively constant average rates of evolution in various phyletic lines during this period (see page 165).

The real situation, however, is more complex. It may be that the process of formation of species is reflected in molecular texts in ways which are not taken into account in the modeling of intrapopulation allele dynamics [76,77].

If constancy is observed in the fixation rate of mutations, it can be attributed to the fact that for tens and hundreds of millions of years, in the absence of serious functional changes from mutations, all the potentially possible positions of the ancestral proteins were included. But the phylogenetic tree is constructed only from observed positional differences, without consideration of the real number of mutations. Even for the most parsimonious tree with respect to the total number of mutations, shown in Fig. 45, in [34a] it is shown that in those globin positions which have varied during the past 300 million years each amino-acid variant has been fixed on the order of two or three times. This is evidence that, for a given protein, the possibilities for the evolutionary selection of amino-acid replacements have to a significant degree been exhausted [34a].

The explanation of the "neutralist paradox" should be sought in the real contents of those positions which have changed, i.e. in the character of the semantic evolution of genes.

In the process of comparing the accumulated mutations in any two lines leading to contemporary species "a" and "b", all those positions which have changed in comparison with the probable ancestral structure can be divided into three groups: 1) positions in which a replacement occurred in line "a" but not in line "b"; 2) positions in which "b" changed, but not "a", and 3) positions in which replacements were fixed in both lines.

It turns out that for the globins of carp and the α-chain of humans, among the positions of the third group there is not one that is involved in a functional center, while in the first two groups such "semantic" positions constitute more than half. Similar results are obtained for any two taxa (at the level of orders or above) which were distributed sufficiently long ago in the evolutionary process. For species which were distributed more recently, as for example, primates, this regularity is not observable, probably due to the too short period of their independent evolution.

Thus the rates of fixation of mutations in various sections of the tree may, on the average, be nearly identical, although the history of each globin chain has its own, frequently unique features, determined by the evolution of functional, semantic positions. Neutral mutations, known to be synonomous, are fixed in "external" noncentral positions.

Thus even in connection with a given gene which is surely not a key one in processes of species formation we clearly fix the specific nature of the species' differences, even on the molecular level, in the semantic section of the molecule. This indicates that the real arena of neutral evolution is much narrower than one would suppose on the basis of the usual constructions.

As a whole, the example of hemoglobins shows that the process of progressive evolution of forms in the direction of more complicated organization is reflected even on the molecular level: The evolution of globins clearly was directed toward ever greater specification of sections of the primary structure (the appearance of new functional centers), i.e. the distinctive enrichment of the semantic content of genetic information within the bounds of the elementary semantic units of genetic systems, the cistrons.

Of course one should bear in mind that the evolution of land organisms is an extremely complex process, depending on many factors; so that it is doubtful whether it took the same course as the hemoglobin system (even on the molecular level). Nevertheless, changes in the hemoglobin genes themselves indisputably played a notable role in it.

Thus we see that the construction of dendrograms for specific families of proteins provides a number of non-trivial pieces of information about the course of macromolecular evolution.

At the present time there is in progress intensive accumulation of factual material, which will permit the consideration of the phylogenetic nearness of various types of proteins. For example, Yčas [100], having established the probable ancestral structures for the most diverse families of proteins (globins, nucleases, lysozymes, cytochromes, etc.), compared them, and was able to identify related families of proteins. This permitted him to conjecture that primordial ancestral amino-acid sequences evolved in different tertiary spatial structures.

Besides the constantly increasing list of deciphered primary structures of the most diverse proteins, information about the structure of polynucleotide texts, tRNA in particular, is multiplying rapidly. The phylogenetic analysis of tRNA is of considerable interest, the goal being to describe the very earliest stages of evolution of genetic systems, since tRNA is a most important part of the translation apparatus, common to all living things. There is ample basis to consider that the noted features of similarity in the structure of tRNA of various fractions from different species shows the commonality of their origin. Therefore, the construction of the tree for tRNA may provide interesting information about the evolution of the genetic code and the translation apparatus [31,52].

In recent years significant progress has been made in deciphering the spatial semantic structure of macromolecules. It can be conjectured that quantitative estimation of the similarity of tertiary structures will permit significantly more complete and specific descriptions of the semantic evolution of genetic texts in the future. It is clear that, as in the analysis of primary structures, the methods of graph theory will play a leading role in this.

Epilogue: Cryptographic Problems in Genetics

From the contents of this book it is clear that the genetic method of investigation is the reconstruction of the structure and properties of the whole by studying its fragments, i.e. it is by its very nature cryptographic.

Cryptographic problems arise in studying any layer of the genetic language. We will recall several of these, both ones which have been solved and some that still await solution.

The first, and most impressive, was the deciphering of the genetic code. This problem was finally solved by direct biochemical experiments. However a problem arose at one stage of the deciphering process: Given the composition of coded words (nucleotide triplets), to find the order of the characters (nucleotides) in them.

One solution was proposed by V A. Ratner. Comparing mutations involving nucleotide replacements in codons and the results of intra-codon recombinations, he was able to establish the order of the nucleotides within codons [29].

At the level of cistrons, the units of translation, at least two cryptographic problems present themselves.

The first of these is the proper reconstruction of the primary structure of cistrons and of the proteins coded by them. There are subtle chemical methods of establishing the nucleotide and amino-acid composition of short fragments of genetic text (5 - 10 nucleotides or amino acids). To establish the structure of longer texts the following method is used. By means of specific proteins (i.e. enzymes such as nucleases and proteinases) the chain being studied is broken down into short, separate fragments. The basic "cryptographic" difficulty is that these fragments are not broken off one after another in sequence, so that to find their ordering it is necessary to repeat the fragmentation process of the initial text, using an enzyme with different specificity. From the resulting large collection of overlapping fragments, it is not difficult to find the desired ordering by hand.

Using this method, up to the present time the primary structures of several hundreds of proteins have been deciphered [58]. However, this is an insignificant fraction of the total number in existence. Moreover the selection of the proteins to be analyzed is usually made on the basis of their amenability to biochemical study, which has little connection with their "value" in a genetic sense.

Frequently genetically interesting texts are not amenable to direct biochemical investigation. However geneticists have available an indirect method for establishing the primary structure of genes and the proteins coded by them, through the construction of recombination maps. In this method individual positions of the primary structure are represented by mutations, and correspondingly the recombination ordering of mutations characterizes the order of these positions in the primary protein structure. Recombination mapping of genes is fundamental to the study of their structure.

With scriptons, we have shown in §1.5 that both recombination and complementation mapping can be used in the reconstruction of their structure.

The problems we have mentioned are associated with the linear organization of genetic systems. However, the most important macromolecular objects of cells are in their very nature three dimensional. Even with individual cistrons, to actualize the information they contain, the linearity of the primary structures is transformed into complex spatial protein structures which are responsible for the functioning of the system. The study of these three-dimensional structures may also be considered as a cryptographic problem, since here also the methods of analysis (both biochemical and genetic) come down to the establishment of the structure and properties of the "whole" by comparing various of its "fragments." The classic example of this is the reconstruction of three-dimensional molecular protein structure from two-dimensional X-ray data (by a map of the electron density of the protein crystals).

Unfortunately, this method is very cumbersome, its basic difficulty being, perhaps, at the first stage, in the obtaining of protein crystals which still have their "active" structure. Therefore the deciphering of the "*physical code*" itself, i.e. the rules of automatic transition from primary to tertiary protein structure, is extremely important[18,26].

Another possible method in the study of proteins is the very significant complementation analysis, the resolving power of which is considered by geneticists to be much weaker than that of recombination mapping. This is understandable, since the complementation test was applied primarily as an express method for determining functional genetic units, and, as it turned out, little could be said about the internal organization of these units. In fact, as we have tried to show in the second chapter, the possibilities associated with this test are considerable. The data obtained from it can be used for quite detailed analysis of functional (semantic) protein organization.

It is important to realize that when a geneticist begins an experiment, as a rule he does not know with what level of complexity of the genetic system he is dealing: individual cistrons, scriptons, operons or more extensive fragments of the genome. Moreover, the estimation of the number of cistrons in a locus is even one of the central problems he must solve.

With this in mind, it is advisable to begin the analysis with the investigation of the macrostructure of the complementation matrix. If it has a clearly discernable linear component, then this is evidence of structural linearity in the system being studied.

It must be said that this approach reflects, in essence, the conjecture already proposed within the framework of classical genetics, that these linear, invariant components of our constructions are images of the real structural topography of genetic loci. The striking ability of genetic systems to retain the linear features of their structural organization throughout a long and tangled history, set the stage for the remarkable achievement of classical genetics in deciphering the structure of the genetic material. Perhaps the most impressive example of this stability of the primary action of genes is the *scute* locus in *Drosophila* (see Section 1.5.3).

The cryptographic problems we have listed are related to the very lowest levels of organization of genetic materials in cells. Of no less interest (and probably of no less complexity) are problems of this type connected with the "top levels", the structural and functional organization of whole blocks of the hereditary program, their grammar and the rules for developing the program in the process of ontogenesis. In this regard, the functional connections of blocks as well as their scheme of coordination clearly have a complex, non-linear character.

For example, by means of "cryptographic" methods of graph theory one can study the "peripheral" portions of intracellular processes, those ramified and sometimes very complex networks of chemical transformation known as the *metabolic processes*. It is useful to mark individual stages in the metabolism process in biochemical genetics with mutations (through changes in the activity of the associated enzymes). Special tests have been worked out (for example, the test for syntrophism or cross-feeding [30,48]), which permit the characterization of paired interactions of mutants. With the matrix of such local connections in hand one can try to reconstruct the whole chain of metabolic reactions.

In addition, already clearly visible is the tempting possibility of applying the language and methods of graph theory in deciphering the genetic control system for ontogenesis. The genetic control mechanisms for ontogenesis are well understood at present only for some viruses (the phages λ, T4, etc.). From graph theory one can expect to construct original at present still "non-structural", but "temporary" portraits characterizing the interactions of the various genes during growth. But the statement of this fundamental problem is a cryptographic one: to construct, from "fragments" (mutational changes in certain stages of a process), the "whole", i.e. the whole of ontogenesis or the individual sequences of ontogenetic events.

Finally, one of the important cryptographic tasks of genetics, for a while yet, evidently, will remain the problem of determining the functional and spatial organization of the entire genome, to which we gave attention in §2.2.

These examples argue persuasively that the methods of graph theory can and must play a notable role in the solution of the most diverse problems of genetics.

Appendix: Some Notions About Graphs

SOME NOTIONS ABOUT GRAPHS

A collection $G = (A, V, \Pi)$ is a *graph* if A is a set of *vertices*, V is a set of links and Π is a (not necessarily everywhere defined) mapping $\Pi: A \times A \longrightarrow V$, which associates with each pair (a,b), for which it is defined, an element $v = \Pi(a,b)$ of the set V — a *link* which connects the vertices a and b in such a way that v does not connect any other pair of vertices, i.e. $v = \Pi(x,y) \longrightarrow \{x,y\} = \{a,b\}$. The mapping Π divides V in a natural way into three parts: V_e — the set of those links $v = \Pi(a,b)$ which join both a with b and b with a, i.e. $v = \Pi(a,b)$ and $v = \Pi(b,a)$; V_a — the set of links $v = \Pi(a,b)$ for which $v \neq \Pi(b,a)$; and V_l — the set of links $v = \Pi(a,b)$ for which $a = b$.

V_a is called the set of *arcs*, V_e the set of *edges*, and V_l the set of *loops*. Each arc, edge or loop is uniquely characterized by the vertices a and b which it connects. In the sequel we will designate an arc (edge) connecting a with b by (a,b) (or more briefly ab or ba, respectively), and a loop at the vertex a by (a,a).

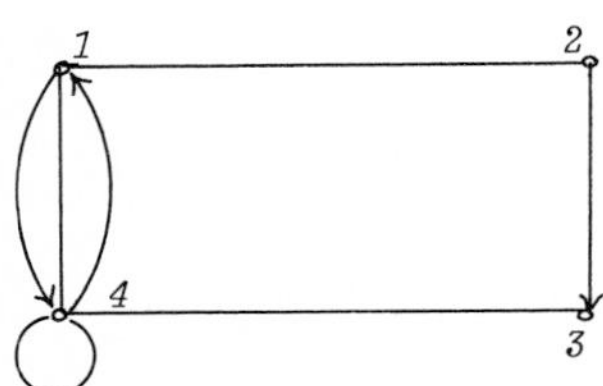

Figure 46

On paper a graph is usually represented as a *diagram*: with *vertices* (elements of A) represented by points, and links $\Pi(a,b)$ by lines connecting the corresponding points a and b. In such diagrams an arc $\Pi(a,b)$ is represented as an arrow from a to b.

For example, the diagram of Fig. 46 corresponds to the graph with $A = \{1, 2, 3, 4\}$, $V = \{v_1, v_2, v_3, v_4, v_5, v_6\}$, in which $\Pi(1,2) = \Pi(2,1) = v_1$, $\Pi(2,3) = v_2$, $\Pi(1,4) = v_3$, $\Pi(4,1) = v_6$, $\Pi(3,4) = \Pi(4,3) = v_4$, and $\Pi(4,4) = v_5$.

Here v_1 and v_4 are edges, v_2, v_3, v_6 are arcs and v_5 is a loop. If $\Pi(a,b)$ is a link, then it is *incident on* the vertices a and b, and such vertices are said to be *adjacent*.

A sequence i_1, $i_1 v_1 i_2$, $i_2 v_2 i_3$, ..., $i_k v_k i_{k+1}$ such that $v_j = \Pi(i_j, i_{j+1})$ for all $j = 1, \ldots, k$ is called a *route* of length k from vertex i_1 to vertex i_{k+1}. In the case when each link of a route is an arc the route is frequently called a *path*, and if all the links are edges it is called a *chain*. Sometimes these conventions are extended to include the case when some of the links are loops, in order to allow the possible repetition of one or more vertices. Agreeable with the above definitions, an individual vertex forms a path whether or not there is a loop at that vertex.

A route (chain, path) is *non-repeating* if all its links are distinct.

The length of the shortest route joining vertices i and j is called the *distance* $d(i,j)$ between i and j. The *radius* of a graph is designated by the number $\min_i \max_j d(i,j)$.

A route $i_1, i_1v_1i_2, \ldots, i_kv_ki_{k+1}$ is called *cyclic* if $i_1 = i_{k+1}$; a cyclic path is called a *contour*; a cyclic non-repeating chain is called a *cycle*.

An *oriented graph* has no edges.

In order to specify an oriented graph it is sufficient to designate those ordered pairs of vertices (a,b) for which Π is defined; an arc $\Pi(a,b)$ being completely specified by the pair $(a\ ,b)$. Consequently an oriented graph is specified by the collection of ordered pairs $(a,b) \in A\times A$ which correspond to its arcs, loops being specified by pairs of the form $(a,a) \in A\times A$. In other words an oriented graph is completely defined by a subset (*binary relation*) $R \subseteq A\times A$ of pairs corresponding to its arcs. Thus in dealing with oriented graphs it is possible and frequently suitable to use the set-theoretic language of binary relations.

Another way to specify oriented graphs is by using *Boolean matrices* — matrices containing only zeros and ones. With a given relation (oriented graph) R is associated a Boolean $N\times N$ matrix $r = \|r_{ij}\|$ of the following form:

$$r_{ij} = \begin{cases} 1, & \text{if } (i,j) \in R, \\ 0, & \text{if } (i,j) \notin R. \end{cases}$$

Clearly such a correspondence of relations and matrices is one-to-one: The relation $R_r = \{(i,j) \mid r_{ij} = 1\}$, on the generated matrix r coinciding with the initial relation R. The matrix r is called the *relation matrix* or *adjacency matrix* of R.

There are, then, three means of specifying oriented graphs (diagrams, relations, Boolean matrices) with extremely simple rules for translating from one set of terms to another. A set of facts can be made more clear by comparing them in these differing terminologies.

The relation $\{(1,2),\ (3,4),\ (2,2),\ (2,3),\ (4,1)\}$ is equivalent to the diagram of Fig. 47 and the matrix

$$\begin{vmatrix} 0 & 1 & 0 & 0 \\ 0 & 1 & 1 & 0 \\ 0 & 0 & 0 & 1 \\ 1 & 0 & 0 & 0 \end{vmatrix}.$$

Also of interest besides oriented graphs are *regular graphs*: graphs whose links are all edges. Sometimes this term is used for non-oriented graphs without loops. Every regular graph can be represented as an oriented graph with the replacement of each edge by two arcs (a,b) and (b,a). Therefore the study of regular graphs may be replaced by the study of corresponding oriented graphs, the "translation" from the language of oriented graphs to that of regular graphs and vice versa being a trivial exercise.

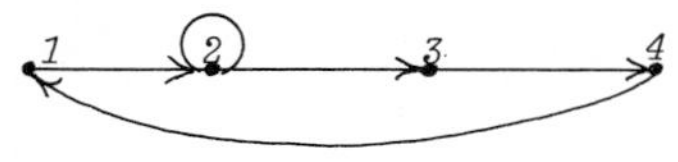

Figure

Thus regular graphs may be represented in terms of relations or Boolean matrices. It is evident that a Boolean matrix corresponds to a regular graph if and only if it is symmetric.

In this book, in the absence of particular stipulations and when it will not be misleading, we will use whichever of these notations (diagrammatic, relational, Boolean-matric) seems appropriate.

For an arbitrary relation $R \subseteq A \times A$, $R\langle a\rangle$ will represent a set, frequently called the *image of the object a with respect to the relation R*, or *the cut (section) of the relation R with respect to the object a*, and defined by

$$R\langle a\rangle \ = \{b \mid (a,b) \varepsilon R\}.$$

In terms of diagrams the set $R\langle a\rangle$ consists of those and only those vertices which are the terminal points of arcs originating at a.

The graph obtained from R by reversing the orientation of its arcs (replacing pairs (a,b) with pairs (b,a)) is said to be *inverse to R* and is designated by $R^{-1} = \{(a,b) \mid (b,a) \varepsilon R\}$. Clearly the matrix of an inverted graph is the transpose of the matrix of the initial graph, i.e. rows are interchanged with columns, the elements r_{ij} being interchanged with r_{ji}.

A *hypergraph* on the set of vertices A is a pair $\Gamma = (A, \{S_i\}_{i \varepsilon I})$, where $\{S_i\}_{i \varepsilon I}$ is a collection of nonempty subsets $S_i \subseteq A$, called *edges* of the hypergraph (*hyperedges*). A hypergraph Γ may be specified by the relation $\gamma \subseteq I \times A$, where $\gamma\langle i\rangle = S_i$, so that $(i,a) \varepsilon\ \gamma \leftrightarrow a \varepsilon S_i$. Such a relation γ (hypergraph Γ) may also be represented by a Boolean $n \times N$ matrix $g = \|g_{ia}\|$, where $n = |I|$ (the cardinality of I) and $N = |A|$ by the rule

$$g_{ia} = \begin{cases} 1, & \text{if } a\ S_i, \\ 0, & \text{if } a\ S_i. \end{cases}$$

The transposed matrix g^T is associated with the inverse relation γ^{-1} and with the inverse hypergraph $\Gamma^{-1} = (I, \{T_a\}_{a \varepsilon A})$ where $i \varepsilon T_a \leftrightarrow a \varepsilon S_i$ as well.

A graph with loops at each vertex corresponds to a relation containing all pairs of the form (i,i) $(i \in A)$ (such a relation is called *reflexive*) and to a matrix with ones along its principal diagonal. Analogously, a graph with no loops corresponds to a matrix with zeros on its principal diagonal, and to a relation containing no pairs of the form (i,i), called an *anti-reflexive* graph.

An oriented graph is *symmetric* if for each arc (a,b) it contains the reverse arc (b,a), in other words, if $R = R^{-1}$. Matrices of symmetric graphs are symmetric.

Oriented graphs corresponding to regular graphs are symmetric.

An oriented graph without loops is *asymmetric* if each link "goes" only one way, i.e. if for each arc (a,b) in the graph, the arc (b,a) is not in the graph. In terms of inverse relations the asymmetry of R means that $R \cap R^{-1} = \phi$. An analogous graph, but with loops, such that $(a,b) \in R$ and $(b,a) \in R \leftrightarrow a = b$, is *antisymmetric*.

Clearly, for a given relation R there exists a unique relation, minimal among those symmetric relations containing R: $R \cup R^{-1}$, obtained by adding all arcs of the form (b,a) for which $(a,b) \in R$. This relation $\tilde{R} = R \cup R^{-1}$ is called the *symmetric closure* of the relation R. In terms of diagrams this operation corresponds to *deorientation* — transforming a graph to a regular graph by the replacement of arcs with edges.

The maximal asymmetric relations contained in a symmetric relation R (there may be many) are obtained by removing from each "edge" $\{(a,b), (b,a)\}$ one of its arcs. For a graph the corresponding operation consists of orienting all edges (replacing them with arcs), and is called the *orientation* of the graph.

A relation (graph) R is *transitive* if, along with the pairs (a,b), (b,c), it contains the pair (a,c). A regular graph is transitive if, along with each pair of edges ab and bc, it contains also the edge ac. In contrast to the preceding situations, these definitions are not entirely consistent: A symmetric oriented graph corresponding to a regular transitive graph satisfies the transitivity condition for $c \neq a$. If c were equal to a then by the definition of transitivity, together with (a,b) and (b,a), the graph would have to contain the loop (a,a), which is not in the original regular graph. For complete transitivity of this graph it is necessary to add loops at all vertices from which an edge extends (i.e. set the diagonal element to one for each nonempty row of the matrix).

This inconsistency between a regular graph and its corresponding graph appears only with loops. In many applications the occurrence or non-occurrence of loops is completely unimportant (we are unconcerned about the diagonal elements of the matrices), so that this inconsistency may be ignored.

Information about the transitivity of a graph can be reformulated as follows. We say that a vertex j is *accessible* from a vertex i if there exitst a path from i to j. A relation is transitive if and only if it coincides with its relation of *accessibilities*. Actually, the definition of trnasitivity means that whatever is accessible in two steps is accessible in one step, and consequently whatever is accessible in three steps is accessible in two steps, etc.

In the general case a graph of accessibilities R does not coincide with the initial graph R. It can be shown that this graph is uniquely minimal among the transitive graphs which contain R; and hence it is called the *transitive closure* of the graph R.

A set S of vertices (objects) is a *clique* in the graph R if $(a,b) \in R$ for any $a,b \in S$, i.e. $S \times S \subseteq R$, or equivalently, if any two vertices of S are joined by arcs in both directions.

A transitive graph has the following important property: The collection of vertices of any contour form a clique, since all vertices lying on the same contour are accessible from one another. In a symmetric graph all the vertices lying on a particular path are mutually accessible, so that through transitivity they also form a clique.

A symmetric transitive reflexive relation is called an *equivalence* relation. In accord with the above, the vertices of an equivalence graph† can be divided into a system of maximal cliques, no one of which is connected with another. If there were an edge joining a vertex of one clique to a vertex of another than all vertices of these two cliques would be mutually accessible, and consequently their union would also form a clique. An equivalence relation R may thus be written in the form

$$R = S_1 \times S_1 \quad \ldots \quad S_m \times S_m, \tag{1}$$

where $S_1, \ldots, S_m$ are maximal cliques which form a *partition of the set* A (they are also called *equivalence classes* of the relation R). The matrix corresponding to an indexed set of objects in the order S_1, $S_2, \ldots, S_m$ reduces to one of block-diagonal form, so that for any $r,t = 1, \ldots, m$, if $i,j \in S_r$ then $r_{ij} = 1$, while $r_{ij} = 0$ for $i \in S_r$, $j \in S_t$ $(r \neq t)$.

The associated partition $S = \{S_1, \ldots, S_m\}$ of the set A is connected in a one-to-one way with an equivalence. To every partition S (i.e. collection of disjoint non-empty subsets covering the set A) there corresponds a relation of the form (1) which is an equivalence relation,

---

† As was remarked earlier, under the transition to a graph, loops (reflexivity) may be disregarded.

in which the classes S_1, ..., S_m of the partition are maximal cliques of the associated graph. This means that yet another language may be used for equivalence relations (graphs) — the language of partitions $S = \{S_1, \ldots, S_m\}$ which, in many applications, has greater clarity than the languages of relations, matrices and diagrams.

In particular, the intersection $P \cap R$ of equivalence relations P and R is itself an equivalence relation with a corresponding partition, the classes of which are non-empty intersections of the classes of the original partitions. The resulting partition is designated by $P \cap R$ and is called the *intersection of the partitions* P and R, where, for convenience we designate partitions by the same letters as their corresponding relations.

Similarly, corresponding to inclusion of equivalence relations $R \subseteq P$ there is inclusion of partitions $R \subseteq P$, meaning that the partition R is more fractional than P, in the sense that each class of P is the union of several classes of R.

Now let R be an arbitrary relation. The following equivalence relations are naturally associated with R. The *connectivity* relation in R, which is the accessibility relation in $R \cup R^{-1}$. Vertices i and j are connected if there is a chain of the non-oriented graph (i.e. where no attention is paid to the orientation of arcs) between them. The transitivity, symmetry and reflexivity of the connectivity relation are evident. The classes of a connectivity relation are called *components* (of connectedness) of the original graph. In other words, components are maximal sets of mutually connected vertices with no connections between different components. A *connected* graph consists of a single component.

The *biconnection relation* is the symmetric part of the accessibility relation. Vertices i and j are *biconnected* if and only if there is a path from i to j or from j to i. That this is an equivalence relation, is evident. The classes of a biconnection relation are called *bicomponents* (*components of strong connectivity*) of the graph. In other words, a bicomponent is a maximal subset of mutually connected vertices.

Vertices i and j are biconnected if and only if they lie on the same contour (characterized by a path from i to j or from j to i), and therefore it is evident that vertices in different bicomponents do not lie on a common contour.

We form a new graph, having as its vertices the bicomponents S_1, ..., S_m in which an arc goes from S_r to S_t if and only if there exists an arc in the original graph from some vertex $i \in S_r$ to some vertex $j \in S_t$. The graph of bicomponents is antisymmetric and contains no contours, and thus, of course, no contours connecting vertices of different bicomponents.

A transitive antisymmetric graph is called a *partial order*. From the foregoing, a graph of the bicomponents of a transitive graph is a partial order.

A *complete* (*linear, perfect*) *graph* has, for every pair of vertices a, b an arc connecting them, i.e. $(a,b) \in R$ or $(b,a) \in R$. A complete partial order is called an *order*: The relation linearly orders the objects so that a path leads from an initial vertex through all vertices in succession. Clearly the graph of bicomponents of a complete transitive relation (often called a *linear quasiorder*) is an order, so that such a graph divides the entire set A into classes (bicomponents), which are its maximal cliques. These classes are linearly ordered, in the sense that from any vertex of a class arcs lead to all those (and only those) vertices which belong to the class or to its successor. It can also be shown that for any ordered partition of the set A there is a corresponding linear quasiorder, the arcs of which are directed from all vertices of preceding classes to all vertices of succeeding classes (and also "internally"). Every partial order is contained in some linear order. For an ordered partition P, the notation aPb means that the class of P which contains a precedes the class containing b.

The relation $R \subseteq A \times A$ is said to be *homomorphic* to the relation $P \subseteq B \times B$ if there exists a single-valued mapping $f: A \rightarrow B$ such that

$$(a,b) \in R \leftrightarrow (f(a), f(b)) \in P.$$

Such a mapping f is called a *homomorphism* from R to P.

Let $a, c \in A$ be such that $f(a) = f(c)$ (for some homomorphism from R to P). Then $R\langle a\rangle = R\langle c\rangle$, for $b \in R\langle a\rangle \leftrightarrow (a,b) \in R \leftrightarrow (f(a), f(b)) \in P \leftrightarrow (f(c), f(b)) \in P \leftrightarrow (c,b) \in R \leftrightarrow b \in R\langle c\rangle$.

Similarly, it can be shown that $R^{-1}\langle a\rangle = R^{-1}\langle c\rangle$.

This leads to the following definition. For a given $R \subseteq A \times A$, a pair of objects a, c are considered *indistinguishable* if $R\langle a\rangle = R\langle c\rangle$ and $R^{-1}\langle a\rangle = R^{-1}\langle c\rangle$. This indistinguishability relation is an equivalence, and the set A is partitioned into non-intersecting classes $R_1, \ldots, R_m$ of indistinguishable objects. On this collection (set) of classes we consider the relation P, consisting of those and only those pairs (R_s, R_t) for which there exist vertices $a \in R_s$, $b \in R_t$ such that $(a,b) \in R$. Clearly, if $(R_s, R_t) \in P$ then for any $c \in R_s$ and $d \in R_t$ the containment $(c,d) \in R$ holds, i.e. $R_s \times R_t \subseteq R$, so that $(a,b) \in R \rightarrow (c,b) \in R \longrightarrow (c,a) \in R$. This means that the mapping $f: A \rightarrow B$, which associates with each object $a \in A$ the equivalence class R_s which contains it, is a homomorphism. This homomorphism and the partition of B are called *canonical*.

It is evident that a relation is an equivalence if and only if it is homomorphic to a graph with loops at each vertex and no other links. Linear quasiorders are similarly characterized by the fact that they are homorphic to a linear order. Indistinguishability relations of such graphs give partitions corresponding to them.

An ordinary graph is called a *tree* if it is connected and has no cycles. There are a number of other characterizations for trees: a) a connected graph with N vertices and $N-1$ edges, b) a graph in which any two vertices are connected by one and only one chain, c) a connected graph which loses that property with the elision of any edge [7].

A type of tree of frequent interest is one in which a particular vertex, called the *root*, is singled out. Such a *rooted tree* is considered to be a *hierarchy*, with the top level containing the root, the second level those vertices directly connected to the root, the third level those vertices connected to vertices in the second level, etc.

A *spanning tree* of a graph R is a tree whose vertex set coincides with the set of vertices of R, and in which each edge is an edge of the graph R. To construct a spanning tree we use the following procedure. Beginning with an arbitrary vertex, include in the tree all edges incident on it which lead to "new," as yet unexamined vertices. If, in this process, no new adjacent vertices are produced, it means that the generated tree is the spanning tree of the selected component of the graph. Now it is necessary to repeat the process, using some, as yet unprocessed vertex. In this way all components of the graph will be selected and their spanning trees generated.

References

1. Andreeva, N. S.: Trehmernaya struktura fermentov. *Žurnal Vsesojuznogo himičeskogo obščestva.* im. D. I. Mendeleeva,16,No.4,1971. (Three-dimensional protein structure.)

1a. Aleskerov, F. T.: *Interval'nyi vybor i evo primenenie v mnogokriterial'nyh zadačah prinjatija rešenii.* Avtoref. dissertacii. M., Institut problem upravlenija,1980,c.17. (*Interval selection and its application in multi-criteria decision making.*)

2. Bačinskii, A. G., Ratner, V. A.: Optimal'nost' i pomehoustoičivost' genetičeskih tekstov. V kn. *Voprosy matematičeskoi genetiki,* ICiG, Novosibirsk,1974,242-261. (Optimality and robustness of genetic texts.)

3. Benzer, S.: Elementarnye ediničy nasledstvennosti. V kn. *Himičeskie osnovy nasledstvennosti,* IL,M.,1960,56. (The elementary units of heredity.)

4. Geršenzon, S. M., Aleksandrov, Ju. N., Maljuta, S. S.: Mutagennoe deistvie DNK i virusov u drozofily. *IMBiG AN USSR,* Naukova dumka, Kiev,1975. (mutagenic action of DNA and viruses in *Drosophila.*)

5. Dubinin, N. P.: Teorija gena, istorija i sovremennye problemy. *Bjull. Mosk. ob-va ispyt. prirody,*69,No.1,1964. (The theory of genes; history and contemporary problems.)

6. Dubinin, N. P., Hvostova, V. V.: Mehanizm obrazovanija složnyh hromosomnyh reorganizacii. *Bio. žurnal,*1935,4,935. (The mechanism of formation of complex chromosomal reorganizations.)

6a. Žarkih, A. A.: Algoritm postroenija filogenetičeskih drev po aminokislotnym posledovatel'nostjam. V sbornike *Matematičeskie modeli evoljučii i selekčii.* (V. A. Ratner, Red.). Novosibirsk, Izd-vo Instituta citologii i genetiki SO AN SSSR,1977,c.5-52. (An algorithm for constructing phylogenetic trees for amino-acid sequences.)

7. Zykov, A. A.: *Teorija konečnyh grafov.* Nauka, Novosibirsk,1969. (*The theory of finite graphs.*)

8. Imrih, V., Stockii, E. D.: Ob optimal'nyh vloženijah metrik v grafy. *DAN SSSR,*1971,200,No.2,279-281. (On the optimal imbeddings of metrics in graphs.)

9. Inge-Večtomov, S. G., Popova, I. A., Gukovskii, D. I., Krivov, V. N.: Rekombinačija i komplementačija v lokuse ad-2 u drozzei *Saccharomyces cerevisiae.* V kn. *Issledovanija po genetike,* Izd-vo LGU,5,1974,35-48. (Recombination and complementation in the locus *ad-2* of the yeasts *Saccharomyces cerevisiae.*)

10. Kemeny, G., Snell, G.: *Kibernetičeskoe modelirovanie.* Sov. radio, M.,1972, (*Cybernetic modeling.*)

11. Kuličkov, V. A., Žimulev, I. F.: Analiz prostranstvennoi organizačii genoma *Drosophila Melanogaster* na osnove dannyh po èktopičeskoi konjugačii politennyh hromosom. *Genetika* ,1976,12,No.5,81-89. (Analysis of the spatial organization in the genome of *Drosophila melanogaster* on the basis of data about the ectopic conuugation of polynemic chromosomes.)

12. Kuperštoh, V. L.: O roli poroga suščestvennosti individual'nyh svjazei v zadače klassifikačii. V sb. 25 . (On the role of an existence threshhold for individual connections in classification problems.)

13. Kuperštoh, V. L., Mirkin, B. G.: Yporjadočenie bzaimosvjazannyh ob"ektov. *Avtomatika i telemehanika,*1971,No.6,77-83; No.7,91-97. (The ordering of interconnected objects.)

14. Kuperštoh, V. L., Trofimov, V. A.: Algoritm analiza struktury matričy sfjazi. *Avtomatika i telemehanika,*1975,No.11,170-180. (An analysis algorithm of the structure of connection matrices.)

15. Kuperštoh, V. L., Mirkin, B. G., Trofimov, V. A.: Summa vnutrennih sfjazei kak pokazatel kačestva klassifikačii. *Avtomatika i telemehanika,*1976,No.3. (The sum of internal connections as an indicator of the quality of a classification.)

16. Kurganov, B. I., Poljanovskii, O. L.: Cetverticnaja struktura i

16. Kurganov, B. I., Poljanovskiĭ, O. L.: Cetvertičnaja struktura i allosteričeskaja reguljačija adtivnosti fermentov. *Ž. Vsesojuznogo him.* ob-va im. D. I. Mendeleeva,1971,16,No.4,421-431. (Quaternary structure and the allosteric regulation of enzyme activity.)
17. Kušev, V. V.: *Mehanizmy genetičeskoi rekombinačii*. Nauka, Leningrad,1971. (*Genetic recombination mechanisms.*)
18. Lim, V. I.: Strukturnye prevraščenija belkovoĭ čepi pri formirovanii nativnoĭ globuly. Gipoteza "izbytočnyh" spiralei. *DAN SSSR*, 1975,222,No.6,1467-1469. (Structural transormations of protein chains in the formation of native globules. The hypothesis of redundant spirals.)
19. Mirkin, B. G.: Ob odnom klasse otnošeniĭ predpočtenija. V sb. *Matem. voprosy formir. èkonom. modelei*. IÈiOPP, Novosibirsk,1970,90-102. (On a class of preference relations.)
20. Mirkin, B. G.: Zadači approksimačii v prostranstve otnošeniĭ i analiz nečislovyh priznakov. *Avtomatika i telemehanika*,1974,No.9, (Approximation problems in a space of relations and the analysis of non-numeric indicators.)
21. Mirkin, B. G.: *Problema gruppovogo vybora*. Nauka,M.,1974. (The problem of group selection.)
22. Mirkin, B. G.: *Analiz kačestvennyh priznakov*. Statistika,M.,1976. (*The analysis of qualitative indicators.*)
23. Mirkin, B. G., Rodin, S. N.: K analizu bulevskih matrič, Svjazannyh s rešeniem nekotoryh genetičeskih zadač. *Kibernetika*,1974,No.2, 108-114. (Toward the analysis of Boolean matrices connected with some genetic problems.)
24. *Problemy analiza diskretnoĭ informačii. C. I*, IÈiOPP,Novosibirsk, 1975. (*Problems in the analysis of discrete information, Part I.*)
25. *Problemy analiza diskretnoĭ informačii. C. II*, IÈiOPP,Novosibirsk, 1976. (*Problems in the analysis of discrete information, Part II.*)
26. Ptičyn, O. B.: Fizičeskie prinčipy samoorganizačii belkovyh šepeĭ. *Yspehi sovrem. biol.*,1970,69,vyp.1,26-48. (physical principles in the self-organization of protein chains.)
27. Ono, S.: *Genetičeskie mehanizmy pregressivnoĭ evoljučii*. Mir,M., 1973. (*The genetic mechanisms of progressive evolution.*)
28. Rao, S. R.: *Lineĭnye statističeskie metody i ih primenenija*.Nauka, M.,1968. (*Linear statistical methods and their applications.*)
29. Ratner, V. A.: *Genetičeskie upravljajuščie sistemy*. Nauka,Novosibirsk,1966. (*Genetic control systems.*)
30. Ratner, V. A.: *Prinčipy organizačii i mehanizmy molekuljarno-genetičeskih pročessov*, Nauka,Novosibirsk,1972. (*Organizational principles and mechanisms of molecular-genetic processes.*)
31. Ratner, V. A.: *Molekuljarno-genetičeskie sistemy apravlenija*. Nauka, Novosibirsk,1975. (*Molecular-genetic control systems.*)
32. Ratner, V. A.: O nekotoryh teoretičeskih problemah molekuljarnoĭ evoljučii. *Zyrn. obščei biol.*,1976,37,No.1,18-29. (On some theoretical problems of molecular evolution.)
33. Ratner, V. A., Furman, D. P., Nikoro, Z. S.: Issledovanie genetičeskoĭ topografii lokusa *scute* y *Drosophila Melanogaster*. *Genetika*, 1969,5,No.6,72-85. (The investigation of the genetic topography in the *scute* locus of *Drosophila melanogaster*.)
34. Ratner, V. A., Rodin, S. N.: O matematičeskoĭ obrabotke matrič allelizma. II: Postroenie i interpretačija drev mežallel'noi komplementačii. V kn. *Voprosy matematičeskoĭ genetiki*, ICiG,Novosibirsk, 1974,214-241. (On the mathematical treatment of allelism matrices, II: The construction and interpretation of interallelic complementation trees.)
34a. Ratner, V.A., Rodin, S. N., Žarkih, A. A.: Issledovanie molekuljarnoĭ filogenii globinov utočnennym metodom. B sbornike *Matematičeskie modeli evoljučii i selekčii* (V. A. Ratner, Red.). Novosibirsk,Ezd-bo Instituta čitologii i genetiki SO AN SSSR,1977,c.53-96. (Investigation of the molecular phylogeny of globins by a more precise method.)
35. Ratner, V. A., Rodin, S. N., Šenderov, A. N.: Problema mežallel'noĭ komplementačii. *Uspehi sovr. biol.*,1975,3,399-419. (The interallelic complementation problem.)

36. Ratner, V. A., Kananjan, G. H.: Postroenie filogenetičeskih drev dlja funkčionalnyh čentrov globinov. V kn. *Issledovanija po matematičeskoĭ genetike*, ICiG, Novosibirsk, 1975, 125-168. (The construction of phylogenetic trees for the functional centers of globins.)
37. Rodin, S. N: Analiz allel'nyh otnošeniĭ recessivnyh letaleĭ, inducirovannyh u drozofily čyžerodnymi DNK i nekotorymi virusami. *Genetika*, 1974, 10, NO.9, 94-105. (The analysis of allelic relations of recessive lethals, induced in *Drosophila* by foreign DNA and several viruses.)
38. Serebrovskiĭ, A. S., Dubinin, N. P.: Iskusstvennoe polučenie mutaciĭ i problema gena. *Uspehi èksper. biol.*, 1929, 4, 235. (Artificial production of mutations and the gene problem.)
39. Soĭdla, T. T.: O strukture lokusa ade-2 drožžei-saharomičetov. *Genetika*, 1972, 8, No.6, 72-80. (On the structure of the *ad-2* locus in saccharomycetes yeasts.)
40. Soĭdla, T. T., Inge-Večtomov, S. G., Simarov, B. V.: Mežallel'naja komplementačija v lokuse AD-2 y drožžei Saccharomyces cerevisiae. *Issledovannija po genetike*, LGU, 1967, 3, 148-164. (Interallelic complementation in the *ad-2* locus of the yeasts *Saccharomyces cerevisiae*.)
41. Stal', F.: *Mehanizmy nasledstvennosti*. Mir, M., 1966. (*Hereditary mechanisms*.)
42. Trofimov, V. A.: K analizu čikličeskoĭ struktury matričy svjazi. V sb. *Voprosy analiza složnyh sistem*, Nauka, Novosibirsk, 1974, 77-83. (Toward the analysis of cyclic structure in connection matrices.)
43. Finčem, Dž.: *Genetičeskaja komplementačija*. Mir, M., 1968. (*Genetic complementation*.)
44. Friedman, G. S.: Nekotorye rezultaty v zadače approksimačii grafov. V kn. [24]. (Some results on the approximation problem for graphs.)
45. Furman, D. P., Rodin, S. N., Ratner, V. A.: Issledovanie genetičeskoĭ topografii lokusa *scute* y *Drosophila melanogaster*. *Soobščenija II-V*, *Genetika*, 1977, Nos.2,4,6. (Investigation of the genetic topography of the *scute* locus in *Drosophila melanogaster*.)
46. Harris, G.: *Osnovy biohimičeskoĭ genetiki čeloveka*. Mir, M.-1973. (*The bases of the biochemical genetics of man.*)
47. Hausdorf, F.: *Teorija Množestv*. ONTI, M.-L., 1937. (*Set theory.*)
48. Heĭs, U.: *Genetika bakteriĭ i bakteriofagov*. Mir, M., 1965. (*The genetics of bacteria and bacteriophages.*)
49. Šaronov, Ju. A., Šaronova, N. A.: Struktura i funkčii gemoglobina. *Molekuljarnaja biologija*, 1975, 9.No.1, 145-172. (The structure and functions of hemoglobin.)
50. Škurba, V. V.: O matematičeskoĭ obrabotke odnogo klassa biohimičeskih èksperimentov. *Kibernetika*, 1965, No.1. (On the mathematical treatment of a class of biochemical experiments.)
51. Šmal'gauzen, I. I.: Što takoe nasledstvennaja informačija? *Problemy kibernetike*, 1966, 16, 23-35. (What is hereditary information.)
52. Eĭgen, M.: *Samoorganizačija materii i evoljučija biologičeskih makromolekul*. Mir, M., 1973. (*Self-organization of matter and the evolution of biological macromolecules.*)
53. Adams, M., Buehner, M., Chandrasekhar, K., Ford, G. C., Hackert, M. L., Liljas, A., Lentz, P., Rao, S. T., Tossmann, M. G., Smiley, I. E., White, J. L., In *Protein-protein interactions* (Ed. by Jaenicke, R., and Helmreich, E.), Springer-Verlag, Berlin, Heidelberg, New York, 1972, 139.
54. Ahmed, A., Case, M. E., Giles, N. H.: The nature of complementation among mutants in the histidine-3 region of *Neurospora crassa*. *Brookhaven Symp. Biol.*, 1964, 17, 53-65.
55. Benzer, S.: On the topology of the genetic fine structure. *Proc. Nat. Acad. Sci.* Wash., 1959, 45, 1607.
55a. Blake, S. S. F.: Exons encode protein functional units. *Nature*, 1979, 277, No.5698, 598.

55b. Carramolino, L., Ruiz-Gomez, M., Guerrero, M., Campuzano, S., Modolell, J.: DNA map of mutations at the *scute* locus of *Drosophila melanogaster*. *EMBO Journal*,1982,1,10,1185-1191.
56. Child, G.: Phenogenetic studies of *scute-1* of *Drosophila melanogaster*. I. The associations between the bristles and effects of genetic modifiers and temperature. *Genetics*,1935,20,N 2,109.
57. Crick, F. H. C.: Orgel, L. E., The theory of inter-allelic complementation. *J. Mol. Biol.*,1964,8,161.
58. Dayhoff, M. O.: *Atlas of protein sequence and structure*, v. 5,Nat. Biom. Res. Foundation, Silver Spring, Maryland, USA,1972.
59. De Serres, F. J.: Carbon dioxide stimulation of the *ad-3* mutants of *Neurospora crassa*. *Mutation Res.*,1966,3,420-425.
60. Fishburn, P. C.: Intransitive individual indifference with unequal indifference intervals. *J. Math. Psychol.*,1970,7,No.1,144-149.
61. Fishburn, P. C.: An interval graph is not a comparability graph. *J. Combinatorial Theory*,1970,8,442-443.
62. Fitch, W. M.: An improved method of testing for evolutional homology. *J. Mol. Biol.*,16,9-16,1966.
63. Fitch W. M.: Toward defining the course of evolution: minimum change for a specific tree topology. *Syst. Zool.*,1971,20,406-416.
64. Fitch, W. M.: A comparison between evolutionary substitutions and variants in human haemoglobins. *Annals of the New York Academy of Sciences*,1974,241,439-448.
64a. Fitch, W. M.: On the problem of discovering the most parsimonious tree. *Amer. Naturalist*,1977,111,No.978,223-257.
65. Fitch, W. M.: Margoliash, E., Construction of phylogenetic trees, *Science*,1967,155,N 4,279.
65a. Flament, C.: Hypergraphs arborés. *Discrete Math.*,1978,21,223-275.
66. Fulkerson, D. R., Gross, O. A.: Incidence matrices and interval graphs. *Pacif. J. Math.*,1965,15,No.3,835-885.
66a. Gilbert, W.: Why genes in pieces? *Nature*,1978,271,501.
67. Gilmore, P. C., Hoffman, J. J.: A characterisation of comparability graphs and of interval graphs. *Canad. J. Mathem.*,1964,
68. Goldschmidt, R.: Die Entwicklungsphysiologische Erklarung des Falls der sogenannten Treppenallelormorphe des Genes *scute* von *Drosophila*. *Biol. Zbl.*,51,507,1930.
69. Goodman, M., Barnabas, J., Maesuda, G., Moore, G. W.: Molecular evolution in the descent of man. *Nature*.,1971,233,No.5322,604-613.
69a. Green, M. M.: Transposable elements in *Drosophila* and other diptera. *Ann. Rev. Genetics*,1980,14,109-120.
70. Greer, J.: Three-dimensional structure of abnormal mutant human haemoglobin. *Cold Spr. Harb. Symp. Quant. Biol.*,1972,36,315.
70a. Hartigan, J. A.: Minimum mutation fits to a given tree. *Biometrics*, 1973,29,53-65.
71. Hartman, P. E., Hartman, Z., Stahl, R. C., Ames, B. N.: Classification and mapping of spontaneous and induced mutations in the histidine operon of *Salmonella*. *Advances in Genetics*,1971,v.16, 1-34.
72. Houston, L. L.: Purification and properties of a mutant bifunctional ensyme from the *hisB* gene of *Salmonella typhimurium*. *J. Biol. Chemistry*,1973,248,No.12,4144-4149.
73. Houston, L. L.: Specialized subregions of the bifunctional *hisB* gene of *Salmonella typhimurium*. *J. Bacteriology*,1973,113,No.1, 82-87.
74. Ingram, V. M.: *The haemoglobins in genetics and evolution*. Columbia Univ. Press,New York,1963.
74a. Karp, R.: Reducibility among combinatorial problems, *Complexity of Computer Computations* (Ed. by R. Miller and J. Thatcher), Plenum, New York,1972,85-103.
75. Kendall, D. G.: Incidence matrices, interval graphs and seriation in archaelogy. *Pacific J. Math.* ,1969,28,No.3,565-570.

76. Kimura, M.: Evolutionary rate at the molecular level. *Nature*,1968, 217,624-626.
77. Kimura, M., Ohta, T.: On some principles governing molecular evolution. *Proc. Nat. Acad. Sci.*,USA,1974,71,No.7,2848-2852.
78. Kuratowski, G.: Sur le probleme des courbes gaushes en topologie. *Fund. Mathem.*,1930,15,271-283.
79. Langley, C. H., Fitch, W. M.: An estimation of the constancy of the rate of molecular evolution. *J. Mol. Evol.*,1974,3,161-178.
80. Lekkerkerker, C. G., Boland, J. Ch.: Representation of a finite graph by a set of intervals on the real line. *Fundam. Math.*,1962, 51,No.1,45-64.
81. Loper, J. C., Grabnar, M., Stahl, R. C., Hartman, Z., Hartman, P. E.: Genes and proteins involved in histidine biosynthesis in *Salmonella*. *Brookhaven Symp. Biol.*,1964,17,15-52.
82. Luce, R.: Semi-orders and a theory of utility discrimination. *Econometrica*,1956,24,178-191.
82a. Maniatis, T., Fritsch, E. F., Lauer, J., Lawn, R. M.: The molecular genetics of human hemoglobins. *Ann. Rev. Genetics*,1980,14,145-178.
83. Michel, J.: An interval graph is a comparability graph. *J. Combin. Theory*,1969,No.2,189-190.
84. Moore, G. W., Gookman, M., Barnabas, J.: An iterative approach from the standpoint of the additive hypothesis to the dendrogram problem posed by molecular data sets. *J. Theor. Biol.*,1973,38,423-457.
85. Moore, G. W., Barnabas, J., Goodman, M.: A method for constructing maximum parsimony ancestral amino acid sequences on a given network. *J. Theor. Biol.*,1973,38,459-485.
86. Perutz, M. F., Lehmann, H.: Molecular pathology of human haemoglobin. *Nature*,1968,219,No.5157,902-909.
87. Perutz, M. F., Muirhead, H., Cox, J. M., Goaman, L. C. G.: Three-dimensional Fourier synthesis of horse oxyhaemoglobin at 2.8 A resolution: the atomic model. *Nature*,1968,219,No.5150,131-139.
88. Ratner, V. A., Rodin, S. N.: Theoretical aspects of genetic complementation. In *Progress in Theoretical Biology* (Ed. by R. Rosen, Snell, F. M.),1976,Acad. Press, v.4,1-63.
89. Roberts, F. S.: Non-transitive indifference. *J. Math. Psychology*, 1970,7,No.2,243-258.
89a. Sakano, H., Huppi, K., Heinrich, G., Tonegawa, S.: Sequences at the somatic recombination sites of immunoglobulin light-chain genes. *Nature*,1979,280,No.5720,288-294.
90. Sankoff, D.: Minimal mutation trees of sequences. *SIAM J. Appl. Math.*,1975,28,No.1,35-42.
91. Scott, D., Suppes, P.: Foundational aspects of theories of measurement. *J. Symbolic Logic.*,1958,23,113-128.
92. Sellers, P. H.: An algorithm for the distance between two finite sequences. *J. Combin. Theory A*,1974,16,No.2,253-258.
92a. Sellers, P. H.: Pattern recognition in genetic sequences. *Proc. Nat. Acad. Sci. U.S.A.*,1979,76,No.7,3041.
93. Simoes Pereira, J. M. S.: On the tree realisation of a distance matrix. *Theorie des graphes*,Actes Journees Int. Etudes ICC,Rome, 1966; Paris, Dunod,1967,383-388.
94. Stern, C.: Genetic mechanisms of development (with localized initiation of differentiation). *Cold Spring Harbor Symp. Quant. Biol.*, 1956,21,375.
95. Stertevant, A. H., Schultz, J.: The inadequacy of the subgene hypothesis of the nature of the scute allelomorphs of *Drosophila*. *Proc. Nat. Acad. Sci. USA*,1931,17,265.
96. Tucker, A.: A structure theorem for the consecutive 1's property. *J. Combin. Theory*,1972,B12,No.2,153-162.
97. Tucker, A.: Matrix characterisations of circular-arc graphs. *Pacif. J. Mathem.*,1971,39,No.2,535-545.
97a. Tucker, A.: Circular arc graphs: new uses and a new algorithm. In *Theory and applications of graphs* (*Proc. Internat. Conf.*,Western

Mich. Univ., Kalamazoo, Mich.,1976),580-589. *Lecture Notes in Math.*, 642,Springer,Berlin,1978.

98. Zuckerkandl, E.: The appearance of new structures and functions in proteins during evolution. *J. Molec. Evol.*,1975,7,1-57.

99. Zuckerkandl, E., Pauling, L.: Evolutionary divergence and convergence in proteins. In *Evolving genes and proteins* (Ed. by V. Bryson and H. J. Vogel),Acad. Press,1965,97-166.

100. Ycas, M.: On certain homologies between proteins. 1976,7,No.3,215-244

Index of Genetics Terms

Index of Mathematical Terms

Biomathematics

Volume 10
A. Okubo

Diffusion and Ecological Problems: Mathematical Models

1980. 114 figures, 6 tables. XIII, 254 pages
ISBN 3-540-09620-5

Contents: Introduction: The Mathematics of Ecological Diffusion. - The Basics of Diffusion. - Passive Diffusion in Ecosystems. - Diffusion of "Smell" and "Taste": Chemical Communication. - Mathematical Treatment of Biological Diffusion. - Some Examples of Animal Diffusion. - The Dynamics of Animal Grouping. - Animal Movements in Home Range. - Patchy Distribution and Diffusion. - Population Dynamics in Temporal and Spatial Domains. - References. - Author Index. - Subject Index.

Volume 9
W. J. Ewens

Mathematical Population Genetics

1979. 4 figures, 17 tables. XII, 325 pages
ISBN 3-540-09577-2

Contents: The Golden Age. - Technicalities and Generalizations. - Discrete Stochastic Models. - Diffusion Theory. - Applications of Diffusion Theory. - Two Loci. - Many Loci. - Molecular Population Genetics. - The Neural Theory. - Generalizations and Conclusions. - Appendices. - References. - Author Index. - Subject Index.

Volume 8
A. T. Winfree

The Geometry of Biological Time

1980. 290 figures. XIV, 530 pages
ISBN 3-540-09373-7

Contents: Introduction. - Circular Logic. - Phase Singularities (Screwy Results of Circular Logic). - The Rules of the Ring. - Ring Populations. - Getting Off the Ring. - Attracting Cycles and Isochrons. - Measuring the Trajectories of a Circadian Clock. - Populations of Attractor Cycle Oscillators. - Excitable Kinetics and Excitable Media. - The Varieties of Phaseless Experience: In Which the Geometrical Orderliness of Rhythmic Organization Breaks Down in Diverse Ways. - The Firefly Machine. - Energy Metabolism in Cells. - The Malonic Acid Reagent ("Sodium Geometrate"). - Electrical Rhythmicity and Excitability in Cell Membranes. - The Aggregation of Slime Mold Amoebae. - Growth and Regeneration. - Arthropod Cuticle. - Pattern Formation in the Fungi. - Circadian Rhythms in General. - The Circadian Clocks of Insect Eclosion. - The Flower of Kalanchoe. - The Cell Mitotic Cycle. - The Female Cycle. - References. - Index of Names. - Index of Subjects.

Volume 7
E. R. Lewis

Network Models in Population Biology

1977. 187 figures. XII, 402 pages
ISBN 3-540-08214-X

Contents: Foundations of Modeling Dynamic Systems. - General Concepts of Population Modeling. - A Network Approach to Population Modeling. - Analysis of Network Models. - Appendices: Probability Arrays, Array Manipulation. Bernoulli Trials in the Binomial Distribution.

Volume 6
D. Smith, N. Keyfitz

Mathematical Demography

Selected Papers
1977. 31 figures. XI, 514 pages
ISBN 3-540-07899-1

Contents: The Life Table. - Stable Population Theory. - Attempts at Prediction and the Theory they Stimulated. - Parameterization and Curve Fitting. - Probability Models of Conception and Birth. - Branching Theory and Other Stochastic Processes. - Cohort and Period, Problem of the Sexes, Sampling.

Springer-Verlag
Berlin
Heidelberg
New York
Tokyo